MUSIC, PASSION, AND COGNITIVE FUNCTION

MUSIC, PASSION, AND COGNITIVE FUNCTION

LEONID PERLOVSKY
Northeastern University, Boston, MA, United States

Academic Press is an imprint of Elsevier
125 London Wall, London EC2Y 5AS, United Kingdom
525 B Street, Suite 1800, San Diego, CA 92101-4495, United States
50 Hampshire Street, 5th Floor, Cambridge, MA 02139, United States
The Boulevard, Langford Lane, Kidlington, Oxford OX5 1GB, United Kingdom

British Library Cataloguing-in-Publication Data
A catalogue record for this book is available from the British Library

Library of Congress Cataloging-in-Publication Data
A catalog record for this book is available from the Library of Congress

ISBN: 978-0-12-809461-7

For Information on all Academic Press publications
visit our website at https://www.elsevier.com/books-and-journals

Publisher: Nikki Levy
Acquisition Editor: Emily Ekle
Editorial Project Manager: Timothy Bennett
Production Project Manager: Julie-Ann Stansfield
Designer: Mark Rogers

Cover image credit: Yakov Vinkovetsky, Metaphysical Composition

CONTENTS

BIOGRAPHY

Dr. Leonid Perlovsky is professor at Northeastern University Psychology Department and CEO LP Information Technology; past Visiting Scholar at Harvard School of Engineering and Applied Sciences, at Harvard Medical School Athinoula Martinos Brain Imaging Center; and Technical Advisor and Principal Research Scientist at the AF Research Lab. He has created a new area of cognitive mathematics, dynamic logic, which models the mind processes. He served as Chief Scientist at Nichols Research, a $500-mm high-tech company; as professor at Novosibirsk University and New York University; and as a principal in commercial start-ups developing tools for text understanding, biotechnology, and financial predictions. His company predicted 9/11 market crash a week before the event and supported the SEC investigation. He is invited as a keynote plenary speaker and tutorial lecturer worldwide, including most prestigious venues such as the Nobel Forum at Karolinska Institutet; published more than 500 papers, 20 book chapters, and 7 books with Oxford, Springer, and Frontiers; and awarded 2 patents. Dr. Perlovsky has founded and serves as Editor-in-Chief for *Physics of Life Reviews* the IF = 9.5, ranked 3 in the world among biophysical journals. He received prestigious awards including the INNS Gabor Award and the John McLucas Award, the highest US Air Force Award for basic research.

ACKNOWLEDGMENTS

This book summarizes breakthrough results in several areas of science. Accomplishing this will not be possible without the help of many colleagues and coauthors. It is my pleasure to acknowledge the contributions of M. Aranovsky, M. Bar, L. Barsalou, M. Bonfeld, R. Brockett, M. Cabanac, A. Cangelosi, T. Chernigovskaya, R. Deming, T. Dudochkin, F. Fontanari, M. Frank-Kamenetskii, J. Gleason, A. Goldwag, D. Huron, R. Ilin, N. Katonova, M. Karpovsky, R. Kozma, L. Leibman, D. Levine, D. Levitin, L. Levitin, T. Lyons, A. Nikolsky, N. Masataka, M. Mazo, Y. Neuman, A. Ovsich, A. Patel, V. Rosenbaum, F. Schoeller, J. Sloboda, D. Sontag, W. Thompson, Y. Vinkovetsky, and B. Weijers for discussions, help, and advice. Development of the ideas in the foundation of this book were supported in part by AFOSR PM Dr. Jon Sjogren, PM Dr. Doug Cochran, and PM Dr. Jun Zhang. Especial acknowledgments are due to Y. Dimitrin who has inspired me to think about the role of musical emotions in human cognition and to D. Vinkovetsky who inspires me to think profoundly.

EPIGRAPH

Art exists so that the bow (of tension between material and spiritual) shall not break.

Nietzsche

A poet's duty is to try to mend
The edges split between the soul and body
The talent's needle. And only voice is thread.

Joseph Brodsky

INTRODUCTION

"Music is an enigma." We spend so much time listening to music. The United States exports more music than guns and cars. Why? Explanations range from bonding of military regiments to dissipating tensions. But these are not explanations at all; they just "pass the buck" to military or psychological uses of music. They do not explain music's ability to touch our souls (what today we call psyche and study in psychology). Anybody who loves music, whether it be Bach or Gregorian chant, The Beatles or Lady Gaga, knows that his or her own love for music cannot be explained by any utilitarian reason. Steven Pinker suggested that "music is auditory cheesecake." He as well as other experts on music, mind, and evolution could not find any biological or evolutionary reason for music. Even Kant, who explained the emotions of the beautiful by relating it to knowledge, could not explain music: "it merely plays with senses," which today is also repeated by Pinker.

So, we are at square one. Why is music so important for us? How is it that rhythms and melodies, just mere sounds, remind states of soul? Is it possible to comprehend the nature of musical perception? I briefly review theories of musical harmony, from Pythagoras to Helmholtz, and show that, while useful for making musical instruments, they do not explain the mystery of music. The mysterious power of music, the content of this book, can only be understood from its relations to the basic mechanisms of the mind: I discuss concepts, emotions, and the knowledge instinct (KI). Music plays a complicated dual role in the workings of the mind. On the one hand, it is perceived by evolutionary-old neural emotional centers connected to ancient instincts. These are mechanisms of the synthetic power of music. On the other hand, music is perceived by evolutionary-new brain center, where music creates new varying and differentiated emotions. These differentiated emotions are required by the KI: for the emotional evaluation of every concept in its multifaceted relationships to our knowledge as a whole. This is the dual role of music: it differentiates emotions and creates synthesis, wholeness in human psyche. Musical sounds engage the human being as a whole—such is the nature of ancient, fuzzy, undifferentiated emotions of the mind. Music at once differentiates emotions and creates wholeness.

In this process of "creating wholeness" music overcomes cognitive dissonances, unpleasant emotions related to contradictions in cognition. People do not like contradictions and avoid them as much as possible, even at the cost of discarding new knowledge. The first description of discarding knowledge to avoid cognitive dissonance was given 2600 years ago by Aesop. In his fable, "The Fox and the Grape," the fox experiences what we call today a cognitive dissonance: the fox sees hanging a beautiful ripe grape, but which is just a little bit too high to reach. The fox cannot get the grape but can avoid suffering from cognitive dissonance; the fox decides "the grape is sour." In the 20th century hundreds of psychological experiments have been used to demonstrate that "if I cannot get it, then I do not need it" as a typical human behavior. And yet humans can overcome cognitive dissonances; otherwise, language, cognition, and culture would not evolve, as every piece of new knowledge contradicts previous knowledge.

This book discusses a revolutionary theory of evolution and the function of music: music helps us to overcome cognitive dissonances, and in this way makes possible the continuous evolution of culture. Furthermore, this theory explains why music has such a power over our souls: knowledge causes grief because every piece of cognition causes cognitive dissonances, we live in this ocean of grief, and music helps us to overcome these negative emotions and to continue living. Is it an exaggeration to call this theory revolutionary? Certainly, conservatories and music schools do not teach anything like this.

There is a fascinating story that empirically supports the theory. The story is about the joint evolution of music and consciousness since the time of King David. This story explains why a particular type of music emerged around the time of Nehemiah, and subsequently entered Christianity. It traces the changes in music and consciousness through Antiquity, the Middle Ages, Renaissance, Reformation, Classicism, Romanticism, the 20th century, fascism, communism, and today. Music through all of this diverse history is related to the interplay of two factors that make up the KI. The evolution of consciousness moves along a razor's edge between differentiation and synthesis: striven to know all of life's details while simultaneously striven to bring unity and meaning to life. Swinging to the side of differentiation, concepts can lose their meaning and purpose; swinging to the side of synthesis, strong emotions can nail down language and thinking to traditional values. At the extreme, both can lead to a cultural slow down. Differentiation is the essence of cultural advancement, but it leads to a breakdown in synthesis. In turn, creative potential is lost and civilization disintegrates. Creative momentum requires a balance between differentiation and synthesis. In the last 2000 years in Western culture the balance has been supported by the evolution of music.

I relate this scientific account to the Nietzschean analysis of the role of music in ancient Greece. Then, I follow cycles of differentiation and synthesis in Western consciousness and music: from early Christian synthesis in psalmody and antiphon, to Medieval differentiation of consciousness and polyphony, to the humanism of the Renaissance and the development of tonality, and to individuation in the Reformation and the fugues of Bach. Why is it that some have called Bach the highest peak in the evolution of music, and yet the evolution of musical styles continues? KI and the evolution of consciousness give us a new point of view on this question. I connect rational consciousness to classical music, romantic consciousness to split between the conceptual and the emotional, and I explore the connections between contemporary consciousness and the music revolutions of the 20th century.

High forms of music create synthesis of the most important concept-representations touching the meaning of human existence. Popular songs, through the interaction of words and sounds, connect the usual words of everyday life with the depths of the unconscious. This is why in contemporary culture, with its tremendous number of differentiated concepts and lack of meaning, such an important role is taken by popular songs. I discuss Rap as a contemporary Dionysian dithyramb, which connects (today, like 3000 years ago) the split edges between the emotional and the conceptual, between soul and body. And I ponder, what is ahead?

I discuss the confirmation of this theory in laboratory experiments. One experiment follows Aesop's fable, but with a slight twist: with music playing in the background knowledge is not discounted. Another experiment demonstrates that during exams music helps students to not avoid cognitive dissonances and to allocate more time to more stressful questions and get better grades. Other researches show that students taking musical classes receive higher grades in all subjects.

The last chapter discusses a new field of study, including the question of measuring musical emotions and their number, and a vast open field for theoretical and experimental exploration.

To summarize, cognitive functions of music, its origins, evolution, and why music has such sway over our souls have been considered a mystery by Aristotle, Darwin, and by contemporary evolutionary musicologists. Kant explained the emotions of the beautiful by connecting them to knowledge, but even Kant could not understand the function of music in cognition. This book presents a revolutionary theory of musical origin based on the fundamental and concrete function of musical emotions in cognition, a theory confirmed by experiments that promises to unify the field and defines new research directions. The book reviews theories of music. It considers the split in the vocalizations of proto-humans into two types: one less emotional and helping to think, evolving into language, and the other preserving ancient emotional connections along with semantic ambiguity, evolving into music. The evolution of language toward the powerful tool of differentiated thinking required freeing the voice from ancient emotional influences. An opposing strive of the human soul demanding synthesis required the compensatory evolution of music toward more differentiated and refined emotionality. The need for pop songs and refined music is grounded in fundamental mechanisms of the mind: evolution of language resulted in the multiplicity of cognitive dissonances, which would stop the evolution of language and culture without the presence of music. This is why today's human mind and cultures cannot exist without today's music. The cognitive functions of music are to overcome cognitive dissonances. We live with cognitive dissonances, which can be unpleasant and even painful. We live in a sea of unpleasant emotions, from minor everyday disappointments, to unrequited love, to the ultimate fear of death, and music helps alleviate this pain. This is why music is so important for us. I discuss empirical traces of this theory in parallel evolution of musical styles and cultures. Experimental verification of this theory in psychological and neuroimaging research is reviewed in the book. Radical differences among musical experts about musical emotions are related to differences in personality types. The final chapter discusses directions of future research and challenges that will need to be overcome.

CHAPTER 1

Theories of Music

Contents

Abstract

Music—why we enjoy it, why it has such power over us, its origins, its evolution, and its cognitive functions—has been considered a mystery by Aristotle, by Darwin, and by contemporary musicologists. This chapter reviews attempts to explain mysterious properties of music from Pythgoras till today. Ancient Greek philosophers saw dangers of untamed emotions and looked for musical appeal to reason. Since Renaissance emotional power of music was gradually accepted, Tonality emerged for expressing emotions in music. Based on Descartes' theory of emotions, "The Doctrine of the Affections" was developed. Deficiencies of Descartes' theory are related to opera seria, Calzabigi and Gluck reform is related to synthesis of emotions and reason. Misunderstanding of cognitive mechanisms affects development of opera. The first attempt at scientific understanding of music begins since Helmholtz, whose theory is briefly summarized. While in the Middle Ages Church has been afraid of strong emotions, today this role is performed by "academic" musicians. Prominent current theories of musical emotions are briefly summarized: while interesting ideas are discussed, scientific understanding is missing, contradictions and confusions abound. Music remains a mystery.

MUSIC IS A MYSTERY

Approximately 2400 years ago Aristotle was the first to ask: "why music being just sounds reminds states of soul?" He listed the power of music among unsolved problems next to finiteness of the world and existence of God. Kant, who so brilliantly explained the epistemology of the beautiful and the sublime, which he related to knowledge, could not explain music: "(As for) the expansion of the faculties ... in the judgment for

Music, Passion, and Cognitive Function.
DOI: http://dx.doi.org/10.1016/B978-0-12-809461-7.00001-7

cognition, music will have the lowest place among (the beautiful arts)... because it merely plays with senses." Darwin discovered laws of evolution of species and thought that everything in the living world can be explained by evolution. There was one exception that Darwin had a great difficulty to explain within his theory: music "must be ranked amongst the most mysterious (abilities) with which (man) is endowed."

Contemporary evolutionary psychologists and musicologists cannot explain music. Pinker follows Kant, suggesting that music is an "auditory cheesecake," a by-product of natural selection that just happened to "tickle the sensitive spots." In 2008, *Nature*, the oldest and most prestigious scientific journal, published a series of essays on music. Their authors agreed that music is a cross-cultural universal, still "none ... has yet been able to answer the fundamental question: why does music have such power over us?" "We might start by accepting that it is fruitless to try to define 'music'." "Music is a human cultural universal that serves no obvious adaptive purpose, making its evolution a puzzle for evolutionary biologists." These are just a sampling of quotes from accomplished scientists.

After reviewing selected theories, I present a new theory based on arguments from cognitive science and mathematical models of the mind suggesting that music serves the most important and concrete function in cognition and in evolution of the mind and cultures. This function is elucidated, its neural mechanisms are discussed, and experimentally testable predictions are made. Then I describe experimental verifications of this hypothesis and discuss parallel coevolution of music and culture.

THEORIES OF MUSICAL EMOTIONS AND MUSIC ORIGINS

During the last two decades, the mysterious powers of music receive scientific foundations due to the research of scientists in several fields. Integration of this research in recent years provides evidence for the evolutionary origins and cognitive functions of music. This chapter provides a selection of views on the function of music in cognition from ancient philosophers to contemporary research.

2500 YEARS OF WESTERN MUSIC AND PRESCIENTIFIC THEORIES (FROM PYTHAGORAS TO THE 18TH CENTURY)

Pythagoras described the main harmonies as whole-number ratios of sound frequencies about 2500 years ago. He saw this as the connection between music and the celestial spheres, which also seemed to be governed by whole numbers. The tremendous potency of music to affect consciousness, to move people's souls and bodies since time immemorial was ambivalently perceived. Ancient Greek philosophers saw the human psyche as prone to dangerous emotional influences and "proper" music served to harmonize the psyche with reason. Plato wrote about the idealized imagined music of the Golden Age of Greece: "... (Musical) types were ... fixed. Afterwards ... an unmusical license set in with the appearance of poets ... men of native genius, but ignorant of what is right and legitimate ... Possessed by a frantic and unhallowed lust for pleasure, they contaminated ... and created a universal confusion of forms ... So the next stage ... will be ... contempt for oaths ... and all religion. The spectacle of the Titanic nature ... is reenacted; man returns to the old condition of a hell of unending misery."

The same appeal to reason among the positive content of music we find 800 years later in Boethius "...what unites the incorporeal existence of reason with the body except a certain harmony, and, as it were, a careful tuning of low and high pitches in such a way that they produce one consonance?" According to foremost thinkers in the 4th and 5th centuries (including St. Augustine), the mind was not strong enough to be reliably in charge of the senses and unconscious urges. Differentiation of emotions was perceived as dangerous.

Only with the beginning of the Renaissance (13–14th century) did the European man feel the power of the rational mind separating from the collective consciousness. For 12 centuries, the positive content of music was seen in its relations to the objective "motion of celestial spheres" and to God-created laws of nature. This changed by the 13th century: Music was now understood as being related to listeners, not to celestial spheres. J. Groceo wrote: Songs for "average people ... relate the deeds of heroes ... the life and martyrdom of various saints, the battles..."; songs for kings and princes "move their souls to audacity and bravery, magnanimity and liberality..." Human emotions, the millennial content of music, were now appreciated theoretically.

Although music had appealed to our emotions since time immemorial, a new and powerful development toward stronger and more diverse emotionality began during the Renaissance. It came with the tonal music that had been developed from the 15th to 19th century with the *conscious* aim of appealing to our musical emotions. Tonality is the system of functional harmonic relations, governing most of Western music. Tonal music is organized around tonic, a privileged key to which the melody returns. Melody leads harmony, and harmony in turn leads melody. A melodic line feels closed, when it comes to rest on (resolved in) tonic. Emotional tension ends and a psychological relaxation is felt in the final move on to the tonic, to a resolution in a "cadence."

Creating emotions was becoming the primary aim of music. Composers strived to imitate speech, the embodiment of the passions of the soul. At the same time the conceptual content of texts increased, "the words (are to be) the mistress of the harmony and not its servant," wrote Monteverdi at the beginning of the 17th century. This became the main slogan of the new epoch of Baroque music. Operatic music was born in Italy at that time.

The nature of emotions became a vital philosophical issue. Descartes attempted a scientific explanation of passions. He rationalized emotions, explaining them as objects and relating them to physiological processes. "Descartes' descriptions of the physiological processes that underlay and determined the passions were extremely suggestive to musicians in search of technical means for analogizing passions in tones."

Based on Descartes' theory, Johann Mattheson formulated a theory of emotions in music, called "The Doctrine of the Affections." Emotions "are the true material of virtue, and virtue is naught but a well-ordered and wisely moderate sentiment." Now the object of musical imitation was no longer speech, the exterior manifestation of emotions, but the emotions themselves.

Beginning from this time musical theory did not just trail musical practice but affected it to a significant extent. Descartes and Mattheson understood emotions as monolithic objects. This simplified understanding of emotions soon led to the deterioration of opera into a collection of airs, each expressing a particular emotion ("opera seria" or serious opera); the Monteverdi vision of opera as integrated text, music, and drama was lost. In the middle of the 18th century Calzabigi and Gluck reformed opera back to the Monteverdi vision and laid a theoretical foundation for the next 150 years of opera development.

As we discuss later, music is different from other arts in that it affects emotions directly, not through concepts as, e.g., visual arts, which first have to be understood conceptually. This clear scientific understanding of the differences between concepts and emotions did not exist during the Renaissance. Nevertheless the idea of music as expression, differentiating and creating new emotions, was consciously formulated in the second half of the 18th century. This idea of music as the expression of emotions led to a fundamental advancement in understanding music as the art differentiating emotions; it related the pleasures of musical sounds to the "meaning" of music. Twining emphasized an aspect of music, which today we would name conceptual indefiniteness: musical contents cannot be adequately expressed in words and do not imitate anything specific. "The notion, that painting, poetry and music are all Arts of Imitation, certainly tends to produce, and has produced, much confusion ... and, instead of producing order and method in our ideas, produce only embarrassment and confusion."

Yet understanding the nature of emotions remained utterly confused: "As far as (music) effect is merely physical, and confined to the ear, it gives a simple original pleasure; it expresses nothing, it refers to nothing; it is no more imitative than... the flavor of pineapple." Twining expresses here a correct intuition (music is not an imitation), but he confuses it with a typical error. Pleasure from musical sounds is not physical and not confined to the ear, as many have thought. As discussed later, pleasure from music is an aesthetic (not bodily) emotion in our mind unlike, e.g., the flavor of a pineapple which promises to our body enjoyment of a physical food. Even the founder of contemporary aesthetics, Kant had no room for music in his theory of the mind. Later we discuss the specific scientific reason preventing Kant from understanding the role of music in cognition. Even today, as discussed later, the cognitive function of musical emotions remain unknown among musicologists; the idea of expression continues to provoke disputes, "embarrassment and confusion."

Whence Beauty in Sound?

A scientific theory of music perception began its development in the first half of the 19th century by Helmholtz's theory of musical emotions, summarized here. A pressed piano key or plucked string produces a sound with many frequencies. In addition to the main frequency *F*, the sound contains overtones or higher frequencies, 2F, 3F, 4F, 5F, 6F, 7F,..., which sound quieter than *F*. The main tone corresponds to the string oscillating as a whole, producing F; on top of this, each part of a string, 1/2, or 1/3 or 2/3..., can oscillate on its own. The interval between *F* and *2F*, double frequency, is called an octave. If *F* is "Do, first octave (256 Hz)," then *2F* is the Do of the second octave.

Our ear almost does not notice an overtone exactly one octave higher, because the eardrum oscillates as a string in concordance with itself (this is the mechanical foundation of musical consonants). For the same reason all exact overtones, $2F$, $3F$, $4F$. . ., are perceived in concordance with the main frequency F and among themselves. Because of the mechanical properties of the eardrum, two sounds with close frequencies (say, F and $0.95F$) produce eardrum oscillations not only with the same frequencies but also with the difference of these frequencies ($F - 0.95F = 0.05F$). These low-frequency oscillations are perceived as physically unpleasant, sounding "rough," and even painful, though at normal loudness they are barely perceived. This is the mechanical reason for musical dissonances. Sounds with exactly the same overtones are perceived as concordant, agreeable, or "mechanically pleasing."

Helmholtz has explained construction of musical scales. He has considered selection of concordant strings within octave, which main overtones equal $3F$, $4F$, $5F$, $6F$, $7F$ (that is the reason for their concordance). These frequencies are above 2F and therefore outside of the first octave. They can be "transformed down" to within the octave, this could be achieved by dividing these frequencies by 2: $3/2F$, $4/2F$, $5/2F$, $6/2F$, $7/2F$. . . (say, by taking a string twice as long). These sounds are perceived by the ear as concordant with the main key (F) and among themselves. This concordance is not as good as among overtones of a single string, but much better than for random sounds. That is the reason for musical importance of the octave: Strings (or keys) separated exactly by an octave (double or half the frequency) have many of the exact same overtones and they sound concordant. Note, only the first of the above sounds, $3/2F$, is within the fist octave (above F and below $2F$); the rest are in the second octave and above. For a key to sound in the first octave and its overtones to coincide with those of Do, we may bring down each overtone by one more octave (or two, or three): $5/4F$, $7/4F$, $9/8F$.

Notes obtained in this way, if we start with the three main overtones, make up the major scale, do, re, mi, fa, sol, la, ti—the white piano keys. They are perceived by the ear as concordant. The note fa, however, sounds more concordant if its first different overtone coincides with an overtone of do, $4F$ (therefore the fa key is chosen as fa $= 4/3F$). Concordance, or similarity of overtones, somewhat depends on the training of the ear, also not all overtones could be made completely concordant; therefore musical acoustics is not as simple as $2 \times 2 = 4$. Musical instruments were improved over thousands of years and they incorporate traditions and compromises. There are important differences among cultures in making musical instruments and tuning them. The most concordant keys do, fa, sol (or *F*, 4/3*F*, 3/2*F*) exist practically in all cultures (they are the most concordant because the first overtone of do is sol, and the first overtone of fa is do). Next four overtones closest in loudness and similarity add up to the major scale.

The minor scale is obtained if the three least concordant keys, mi, la, ti, are lowered by a half-tone (tone = 1/7th of an octave), so that they are more concordant with the other less loud overtones. If one chooses just one the most concordant note among these three less concordant keys, the note la, the resultant 5-notes are called the pentatonic scale; it is used in Chinese music, in folk music of Scotland, Ireland, and in Africa.

The scale of an accurately tuned piano slightly differs from what is described above. The reason is that all overtones of all keys cannot coincide; scale based on overtones of do is not as well concordant with overtones of other keys. For example, an overtone of mi, similar to sol, is 1/4 tone different from sol and sounds as a strong dissonance. For string instruments, such as a violin, it is not too important; a violinist can take the correct interval for each note, similarly a singer. But for keyboard instruments, like piano, this sound error is not correctable. Therefore, in the 16th century a well-tempered scale was developed, which divides an octave into 12 equal intervals (half-tones), so that errors in the main overtones are equally spread and all keys are slightly discordant. Concordant musical sounds are called consonances, and less concordant, dissonances. The exact meanings of these words change with culture.

The Helmholtz acoustic theory explained the role of the octave, tones making up the octave, the major scale (do, re, mi, fa, sol, la, ti—the white piano keys) and the minor scale; the pentatonic scale. Notwithstanding Helmholtz's theory, there is a principled difference between the "mechanical" agreeableness of concordant overtones and the aesthetic beauty of music. For example, the minor scale is aesthetically interesting exactly due to its slight discordance. Therefore Helmholtz's theory could not be accepted as a basis for musicology. Acoustic properties of the human voice and ear do not guarantee that Mozart sounds "naturally." A single string sounds naturally in complete concordance with its overtones, but classical musical harmony used natural mechanisms of perception of consonances and dissonances for complex aesthetic effects. The fundamental significance of Helmholtz's theory remained unclear because it was not connected to the aesthetic meaning of music.

Recent laboratory experiments have confirmed that musical harmony is based on inborn mechanisms. Babies, 2 days of age, prefer consonant sounds over dissonances. Evolution, it seems, used mechanical properties of the ear for enhancing the efficiency of the voice communication channel. As a string made of inhomogeneous material sounds in discordance with itself, so does the human voice chord, when in stress or fear it sounds discordant; and this discordance was perceived as unpleasant millions of years ago. In the basis of human voice communication, there are consonant combinations of sounds. These have been gradually evolving into the emotionally filled melody of voice. The connection of voice sounds with the states of soul was inherent in our ancestors long before language began evolving toward conceptual content at the expense of the emotional one. Gradually, evolution shaped our musical ability to create and perceive sound as something principally important, touching all of our being.

Another physical difficulty of Helmholtz's theory is that the emotional perceptions of consonances and dissonances extend from contemporaneously sounding frequencies also to temporal sequences of tones, and this cannot be explained by beats of the eardrum. Apparently, over millennia (or possibly over millions of years beginning in animals—this point might be contentious) neural mechanisms added to our perception of the originally mechanical properties of the ear. I'll add that Helmholtz did not touch the main question of why music is so important psychologically—this remained a mystery.

Current Theories of Musical Emotions

Most existing theories contradict each other and cannot explain mechanisms or functions of musical emotions in workings of the mind; cannot explain evolutionary reasons for music origins, or why it is so important for us. Music still seems to be an enigma. Current theories of musical emotions attempt to uncover this mystery by looking into its evolutionary origins. Justus and Hutsler, and McDermott and Houser review evidence for the evolutionary origins of music. They emphasize that an unambiguous identification of genetic evolution as a source of music origins requires innateness, domain specificity for music, and uniqueness to humans (since no other animals makes music in the sense humans do). The conclusions of both reviews are similar, i.e., "humans have an innate drive to make and enjoy music." There is much suggestive evidence supporting a biological predisposition for music. Certain basic abilities for music are guided by innate constraints.

Still, it is unclear that these constraints are uniquely human since they "show parallels in other domains." It is likely that many musical abilities are not adaptations for music, but are based on more general-purpose mechanisms. "Available evidence suggests that the innate constraints in music are not specific to that domain, making it unclear, which domain(s) provided the relevant selection pressures." "There is no compelling reason to argue categorically that music is a cognitive domain that has been shaped by natural selection." In *Nature*'s series of essays on music McDermott writes: "Music is universal, a significant feature of every known culture, and yet does not serve an obvious, uncontroversial function." Darwin indeed had strong reasons for considering music a unique mystery.

In commentaries to these reviews, Trainor argues that for higher cognitive functions, such as music, it is difficult to differentiate between adaptation and exaptation (structures originally evolved for other purposes and used today for music), since most such functions involve both "genes and experience." Therefore the verdict on whether music is an evolutionary adaptation should be decided based on its advantages for survival. In a recent publication Trainor argues that music originally evolved for auditory scene analysis and concludes that the origins of music are complex. Fitch comments that the biological and cultural aspects of music are hopelessly entangled, and "the greatest value of an evolutionary perspective may be to provide a theoretical framework."

Before reviewing other select authors, I would comment that the theory presented later in this book corresponds to many of the suggestions and ideas in this chapter. In addition, I discuss a concrete and fundamental function of musical emotions in cognition and the evolution of language, mind, and culture, which is missing in other theories and which provides new directions to search for the evolutionary mechanisms of music. This book relates to the biological roots of music, to its origins in "an earlier system of affective communication," it bears on the discussions of evolution versus exaptation, and human symbolic abilities. Possibly, most important it clearly explains why many people are so strongly affected by music.

Huron emphasizes that in the search for the evolutionary origins of music it is necessary to look for complex multistage adaptations, built on prior adaptations, which might have evolved for several reasons. He lists several possible evolutionary advantages of music: mate selection, social cohesion, the coordination of group work, developing auditory skills, refined motor coordination, conflict reduction, preserving stories of tribal origins. Nevertheless emphasizes Huron, the list of possible uses of music by itself does not explain musical power over the human psyche; does not explain why music and not some other, nonmusical activities have been used for these purposes. Huron proposes his theory of musical emotions based on psychology of expectations; while psychologically interesting, it does not relate musical emotions to fundamental mind mechanisms.

Antonio Damasio discusses two opposite approaches to understanding emotions in music, which today are called emotivist and cognitivist. Emotivists posit that people really feel emotions when listening to music, whereas cognitivists suggest that feelings of emotions in music are based on first, understanding of the emotional content in music, which follows by emotions stimulated by this cognitive content. This opposition between emotivists and cognitivists is important for understanding many contradictions in music cognition. Later in Chapter 8, Musical Emotions and Personality, after connecting pleasure of music to fundamental cognitive mechanisms, I discuss the psychological reason for this controversy between emotivist and cognitivist understanding of music and relate it to differences in human psychological types.

A cognitivist approach sometimes is considered to originate from the 19th century musical critic Eduard Hanslick, who has thought that "the meaning of music is the form of music." There are no meanings in music, he wrote, outside of its form, in particularly he criticized appreciation of music based on feelings, which of course is the way most of people appreciate music. Certainly the Hanslick idea that the meaning of music exists independently of humans does not lead to explaining cognitive functions or the origin of music. This extreme one-sided view of the opposition discussed by Damasio might seem absurd, still it widely influences musical profession until today, e.g., Kivy. Therefore relating these extreme views to psychological types in Chapter 8, Musical Emotions and Personality, is important for understanding psychological basis for theoretical controversies.

Let us move to contemporary theories of music origins. Ian Cross, Cross and Morley concentrate on evolutionary arguments specific to music. Cross emphasizes that it is inadequate to consider music as "patterns of sounds" used by individuals for hedonic purposes. Music should be considered in the context of its uses in precultural societies for social structuring, forming bonds, and group identities. A strong argument for the evolutionary origins of music is its universality; music exists in all scientifically documented societies around the globe. Cross emphasizes that music possesses common attributes across cultures: it exploits the human capacity to entrain to social stimuli. He argues that music is necessary for the very development of culture. Cultural evolution is based on the ability to create and perceive the sociointentional aspect of meaning. This is unique to humans and it is created by music. Cross presents a three-dimensional account of meaning in music, combining "biologically generic, humanly specific, and culturally enactive dimensions." Thus the evolution of music was based on biological and genetic mechanisms already existing in the animal world.

The capacity for culture requires not only the transmission of information but also the context of communication. Therefore "music and language constitute complementary components of the human communicative toolkit." The power of language is in "its ability to present semantically decomposable propositions." Language, because of its concreteness, on one hand enabled the exchange of specific and complicated knowledge, but on the other hand could exacerbate the oppositions between individual goals and transform an uncertain encounter into a conflict.

Music is a communicative tool with opposite properties. It is semantic, but in a different way than language. Music is directed at increasing a sense of "shared intentionality." Music's major role is social, it serves as an "honest signal" (that is, it "reveals qualities of a signaller to a receiver") with nonspecific goals. This property of music, "the indeterminacy of meaning or floating intentionality," allows for individual interactions while maintaining different "goals and meanings" that may conflict. Thus music "promotes the alignment of participants' sense of goals." Therefore Cross hypothesized that successful living in societies promoted the evolution of such a communication system.

Cross suggests that music evolved together with language rather than as its precursor. Evolution of language required a rewiring of neural control over the vocal tract, and this control had to become more voluntary for language. At the same time a less voluntary control, originating in ancient emotional brain regions, had to be maintained for music to continue playing the role of the "honest signal." The related differences in neural controls over the vocal tract between primates and humans were reviewed by Perlovsky and will be summarized later.

As juvenile periods in hominid lineages lengthened, so-called altricialization, music took a more important role in social life. The reason is that juvenile animals, especially social primates, engage in play, which prepares them for adult lives. Play involves music-like features, thus proto-musical activity has ancient genetic roots. Lengthening of juvenile periods was identified as being possibly fundamental for proto-musical activity and for the origin of music. Infant-directed speech, IDS, has special musical (or proto-musical) qualities that are universal around the globe. This research was reviewed by Sandra Trehub. She has demonstrated that IDS exhibits many similar features across different cultures. Young infants are sensitive to musical structures in the human voice. Several researchers relate this sensitivity to the "coregulation of affect by parent and child," and consider IDS to be an important evolutionary mechanism of music origin. Yet the arguments presented later suggest that IDS cannot be the full story behind musical evolution. Recently Honing supported IDS as a contributing factor in the origin of music.

Ellen Dissanayake considers music primarily as a behavioral and motivational capacity. Naturally evolving processes led to the ritualization of music through formalization, repetition, exaggeration, and elaboration. Ritualization led to arousal and the shaping of emotions. This occurred naturally in IDS through the process of mother–infant interaction, which in addition to the specially altered voice involved exaggerated facial expressions and body movements in intimate one-to-one interaction. Infants 8 weeks old already are sensitive to this type of behavior, which reinforces emotional bonding. This type of behavior and the infants' sensitivity to it are universal throughout societies, which suggests an evolved inborn predisposition. Dissanayake further emphasizes that such proto-musical behavior has served as the basis for culture-specific inventions of ritual ceremonies for uniting groups as they united mother–infant pairs. The origins of music, she emphasizes, are multimodal, involving aural, visual, and kinesic activity. She describes structural and functional resemblances between mother–infant interactions, ceremonial rituals, and adult courtship, and relates these to the properties of music. All of these, she proposes, suggest an evolved "amodal (independent of senses—aural or visual) neural propensity in the human species to respond—cognitively and emotionally—to dynamic temporal patterns produced by other humans in the context of affiliation."

This combination of related adaptations was biologically motivated by the cooccurrence of bipedalism, expanding brain size, and altricialization and was fundamental to human survival. This is why, according to

Dissanayake, proto-musical behavior produces such strong emotions and activates brain areas involved in ancient mechanisms of reward and motivation, the same areas that are involved in the satisfaction of our most powerful instincts: hunger and sex. I am not convinced by this argument: hunger and sex are common among all higher animals, but music is unique to human.

Mithen argues that Neanderthals may have had proto-musical ability, that music and language have evolved by the differentiation of early proto-human vocal sounds like "Hmmmm" and undifferentiated proto-music-language. The development was facilitated by vertical posture and walking, which required sophisticated sensorimotor control, a sense of rhythm, and possibly the ability for dancing.

The differentiation of Hmmmm, he dates to after 50,000 BCE. Further evolution toward music occurred for religious purposes, which he identifies with supernatural beings. Currently music is not needed, it has been replaced by language, it only exists as inertia, as a difficult-to-get-rid-of remnant of the primordial Hmmmm. An exception could be religious practice, where music is needed since we do not know how to communicate with God. I don't believe in dismissing Bach, Beethoven, or Shostakovich in this way. Also the implied characterization of religion does not make much sense, why cannot we communicate with God using language, and why would music help? This suggestion also contradicts the history of religions.

Mithen summarizes the state of knowledge about vocalization by apes and monkeys. Unlike older views, he states that calls could be deliberate, however their emotional-behavioral meanings are not differentiated; this is why primates cannot use vocalization separately from emotional-behavioral situations (and therefore cannot develop language), this area is still poorly understood. While addressing language in details, Mithen (and other scientists as well) give no explanation for why humans learn language by about the age of 5 years, but the corresponding mastery of cognition (understanding events in the world) takes a lifetime; steps toward explaining this are taken by Perlovsky and summarized in this book; it is an essential part of the theory of music.

Mithen's view on religion contradicts the documented evidence for the relatively late proliferation of supernatural beings in religious practice, and to cognitive-mathematical explanations for the role of the religiously sublime in the workings of the mind, which is discussed later in the book.

Juslin and Västfjäll and Juslin analyze the mechanisms of musical emotions. They emphasize that in the multiplicity of reviews considering music and emotions, the very use of the word "emotion" is not well defined. They discuss a number of neural mechanisms involved with emotions and different meanings implied for the word "emotion." I would mention here just three of these. First, consider the so-called basic emotions, which are most often discussed; we have specific words for these emotions: fear, sexual-love, jealousy, thirst, and others. The mechanisms behind these emotions are related to the satisfaction or dissatisfaction of basic instinctual bodily needs such as survival, procreation, and a need for water balance in the body. The ability of music to express basic emotions is a separate field of study, not related to ability for music and not touched in this book. Juslin adds aesthetic emotions as an important type of emotions in his theory of musical emotions; however no definition of aesthetic emotions is given by Juslin, no by other authors. Some authors consider the complex or "musical" emotions (sometimes called "continuous"), which we "hear" in music and for which we do not necessarily have special words. This book defines aesthetic emotions, discusses their specific functions in cognition and in which way they are different from basic emotions; musical emotions are aesthetic emotions, their mechanisms, functions in the mind, and cultural evolution are the main subjects of this book.

Levitin classified music in six different types, fulfilling six fundamental needs and (as far as I understood him) eliciting six basic emotions. He suggests that music originated from animal cries and it functions today essentially in the same way, communicating emotions. Emotions motivate us to act: "emotions and motivation are two sides of the same evolutionary coin." It is more difficult, he writes, "to fake sincerity in music than in spoken language." The reason that music evolved this way as an "honest signal" was because it "simply" coevolved with the brain "precisely to preserve this property." I disagree: even nonhuman animals can fake their cries: seagulls when seeing food cry "danger" in their "language" to scare competitors. So it does not seem that music "simply" coevolved for preserving ability for "honest signal"; just think about actors, singers, and poets, contemporary professionals as well as those in traditional societies existing since time immemorial. Trehub also opposes this "honest signal" argument and calls music a "dishonest signal."

I would mention analysis of musical acoustic and psychological content, which is separate from the question of music origin and evolution, and still of great related interest. A classical work of L. Meyer analyzed the meaning in music in terms of old oppositions between intellect and emotions: absolutist and formalist versus referentialist and expressionist. The first pair of terms refers to the meaning of music exclusively within the music itself (following Hanslick) and to be primarily intellectual. The second pair of terms refers to the meaning of music in its reference to human emotions. Even so the opposition between intellect and emotions is to some extent outdated, still it influences many opinions today. Later in the book, in Chapter 8, Musical Emotions and Personality, I relate strong effects of these old oppositions on music perception to psychological types: people indeed are different, some are more conceptual, other are more emotional, and these oppositions are peculiarly interwoven in our conscious and unconscious selves.

Mathematical modeling of acoustic features of music perception and musical emotions was considered by Purwins, Cangelosi, and others. These approaches use low-level psychoacoustic musical features for predicting valence (pleasant—unpleasant) and arousal of musical emotions. A broader attempt to measure richness of musical emotions is GEMS scale, which uses up to 45 emotion labels. It is a significant advance comparative to two-dimensional models of emotions.

Yet GEMS are still limited to language, to emotions for which there are specific words. According to the theory described in this book it is inadequate in principle, music has evolved specifically to overcome limitations of language, for understanding and expressing emotions. There are approximately 150 emotional words in English, and according to Petrov et al these 150 words express only few different emotions (various authors estimate between 2 and 20, depending on the choice of criteria for "different"). As discussed later, this is grossly inadequate for thousands of musical emotions that many of us hear in music of every significant composer.

Let us return for a moment to the beginning of this book and a quote from Darwin that music is the greatest mystery. Along this suggestion Darwin hypothesized that the only way for resolving this mystery he saw if music could be a result of sexual selection, similar to a peacock tail: if music ability is attractive to females and selected by their choice. Darwin apparently had great doubts in this idea: he devoted to it less than two pages in his thousand page book. Since Darwin the idea of sexual selection for music has often been discussed, yet no confirmation has been obtained. In a recent research, on the opposite, contradictory results have been received. The music is not a peacock tail.

Difficulties Remain. Music Is a Mystery

Discussions of the mechanisms that evolved music from IDS to Bach and The Beatles in previously proposed theories are lacking or unconvincing.

Why do we need the virtual infinity of "musical emotions" that we hear in music, e.g., in classical Western music as well as in pop songs? Is it an aberration or do these emotions address potentially universal human needs? Dissanayake suggests that this path went through ceremonial ritualization, due to "a basic motivation to achieve some level of control over events..." Dissanayake appreciates that this line of arguments cannot explain how proto-human IDS evolved into Palestrina, Bach, and Beatles. Eventually she recluses herself from explaining origin of music: If "for five or even ten centuries ... music has been emancipated from its two-million year history ... its adaptive roots says more about the recency and aberrance of modernity ..." Cross and Morley argue against this conclusion: "... it would be impossible to remove music without removing many of the abilities of social cognition that are fundamental to being human." Cross concludes that "there are further facets to the evolutionary story (of the origins of music) requiring consideration. Investigation of the origins, emergence, and nature of musical behaviours in humans is in its early stages, and has plenty more to reveal."

In absence of understanding of any fundamental cognitive functions of music, some musicologists are searching for origins across species, others are searching for social origins of music. In 2015, distinguished cognitive musicologists summarized the current state of understanding the mystery of music in a special issue of *Philosophical Transaction*. "Why do we have music? What is music for, and why does every human culture have it? Is it a uniquely human capability, as language is?" Their conclusions are that to answer these questions it is important to identify human and nonhuman mechanisms of musicality, their functions. "It is virtually impossible to underpin the evolutionary role of musicality as a whole." The fundamental questions about music remain unanswered. Music still is a mystery.

In the following chapters the new theory answers these fundamental questions about music and provides bases for further research in many directions. I discuss a most important and concrete function of music in cognition, which makes the entire cultural evolution possible, explains evolution of music from the differentiation of original proto-music-language to its contemporary refined states, and explains why music "just mere sounds" "reminds the states of soul." This theory has made predictions that have been experimentally confirmed. I would add that no other theory makes nontrivial experimentally verifiable predictions. Of course, since Newton, making predictions and verifying them in experiments is the hallmark of science.

CHAPTER 2

Mechanisms of the Mind: From Instincts to Beauty

Contents

Abstract

Fundamental mechanisms of the mind are discussed as steps to understanding music: concepts, instincts, and emotions. Aesthetic emotions are related to the knowledge instinct. The top of the mind hierarchy is analyzed: emotions of the beautiful are related to the understanding of the highest meaning and purpose.

CONCEPTS, INSTINCTS, EMOTIONS, AND BEHAVIOR

This chapter summarizes the fundamental mechanisms of the mind: concepts, instincts, emotions, and behavior; these serve as a first step toward more complicated mechanisms essential for understanding the role and evolution of music. The content of this chapter summarizes a neurocognitive and mathematical theory developed in many publications.

The mind understands the world in terms of concepts. *Concepts* are neural representations that operate as mental models of objects and events. This analogy is quite literal, e.g., during visual perception of an object, a concept-model in the mind projects an image onto the visual cortex, which is matched there to an image projected from the retina (this simplified description is discussed in more details in the references). Proof of this mechanism in experimental neuroimaging, including detailed descriptions of the brain regions involved was obtained by Bar et al. Perception occurs when the two images are successfully matched.

Instincts are mechanisms of survival that are much more ancient than mechanisms of concepts. Psychological literature actively discusses mechanisms of instincts and these discussions can be followed in the given references. Here we follow these references in considering the mechanism

Music, Passion, and Cognitive Function.
DOI: http://dx.doi.org/10.1016/B978-0-12-809461-7.00002-9

of instincts as similar to the internal sensors that measure vital organism parameters important for normal functioning and survival. For example, a low sugar level in the blood indicates an instinctual need for food. This sensor measurement and the requirement to maintain sugar level within certain limits is a mechanism of "instinct." The function of satisfying instinct is an appropriate level of analysis for this chapter (more details can be found in the literature).

Emotions designate a number of various mechanisms which are surveyed in various publications. In this book the mechanism of emotions are neural signals connecting instinctual and conceptual brain regions. Emotions, emotional neural signals, related states, and feelings communicate instinctual needs to conceptual recognition—understanding mechanisms, so that concept-models corresponding to objects or events that can potentially satisfy instinctual needs receive preferential attention and processing resources within the brain. Thus emotions evaluate concepts for the purpose of instinct satisfaction. Emotional signals and related states of the mind are felt as emotional feelings.

Contemporary psychological research of emotions is usually limited to basic emotions, which are named by words, related to satisfaction of bodily instinctual needs and limited in number to a few different emotions. This book emphasizes that these few basic emotions are a tiny part of our emotional abilities, although the most ancient and salient ones. Our higher cognitive abilities involve a virtual infinity of "continuous" emotions, which are not described by specific words and include emotions in the prosody of voice, emotions of cognitive dissonances, which are briefly described later, as well as musical emotions, the main topic of this book.

Conceptual-emotional understanding of the world results in bodily actions and actions within the mind. Most of *behavior* occurs in the mind. This is the behavior of improving understanding and knowledge, the behavior inside the mind directed at improving concepts. Let us mention that there are "lower level" autonomous behavioral responses, which humans share with animals and which do not involve mechanisms of concepts. We do not need to consider them here for understanding the function of music.

The above theory describing conceptual-emotional recognition and understanding encompasses the mechanisms of intuition, imagination, planning, conscious, unconscious, and many others, including aesthetic emotions. Most brain operations are unconscious, e.g., individual neuronal firings usually can never be accessed by consciousness. This book refers to the brain's neural processes that are not accessible to consciousness as being unconscious, and there are various degrees of unconsciousness. Some processes could never become conscious; others can be accessed by consciousness with significant mental effort, as in creative processes; still others become conscious under changing circumstances without special effort.

Here we touch on one mechanism of the mind, which is a part of perception and imagination. For example, visual imagination occurs when one contemplates objects or situations with closed eyes. Contemplated concept-models project images onto the visual cortex causing visual imagination. A significant part of perception is an unconscious process; e.g., visual perception takes about 150 ms, which is a long time when measured in neuronal firings: about 10 ms per neuron, while tens of thousands of neurons are participating in parallel. The initial part of this process cannot be accessed by consciousness. The initial concept-model projections from memory onto the visual cortex are vague and the human mind is not conscious of them. Only when concept-model projections match object-projections from the retina and become crisp do conscious perceptions occur. It is possible to make the vague concept-model projections conscious: close your eyes and imagine an object in front of you; this imagination is usually vague, not as crisp as the perception of an object with eyes open. This vagueness of imaginations testifies to the vagueness of concept-models. This vagueness is essential for understanding the working of the mind and ultimately the cognitive function of musical emotions. Let us now move to the mechanisms of aesthetic emotions, a special class of emotions including musical emotions, related to knowledge.

THE KNOWLEDGE INSTINCT AND AESTHETIC EMOTIONS

To satisfy instinctual needs, e.g., eating or procreation, the mind must first perceive a variety of surrounding objects as well as understand specific situations. This requires the mind matching concept-models to the surroundings, mostly projected to the visual cortex. But the surrounding objects never exactly match old concept-model memories. Angles, lighting, positions of the objects, and surrounding objects are always different. This has presented difficulties in developing artificial intelligence and pattern recognition systems since the 1950s, with significant progress made only recently. To overcome these difficulties of complexity even in simplest visual scenes, the initial projections of concept-models are vague and they approximately match many different objects. To actually perceive specific objects, the mind has to modify concepts so that they "fit" concrete objects and situations observed. This mechanism operates independently of our desire "to perceive," it is an inborn autonomous mechanism, more fundamental than eating or love (because eating requires seeing the food). It is aimed at satisfying a basic need to understand the world around us by making concept-models "similar" to the surroundings. The mind has an inborn instinct that "senses" this similarity and maximizes it. This mechanism is called the knowledge instinct. Knowledge is the measure of correspondence between concepts and the world.

Emotions that evaluate satisfaction or dissatisfaction of this instinct are felt as harmony or disharmony between the knowledge and the world. They are not related directly to "lower" bodily needs, but only to "higher" need for knowledge. In this sense they are "higher," "spiritual," *aesthetic* emotions (emotions related to knowledge are called aesthetic since Kant). It is in this way that Kant explained the emotion of the beautiful, which I discuss later. Kant could not complete his explanation to his satisfaction, because he did not consider the need to constantly adapt concept-models, he did not know about the knowledge instinct. I would like to emphasize that aesthetic emotions are not peculiar to the perception of art; they are inseparable from every act of perception and cognition. The relation of these emotions to the beautiful and to musical emotions is considered later. During the perception of everyday objects, these emotions usually are below the threshold of conscious registration. We do not feel emotionally elated when correctly understand, say a refrigerator, or another simple everyday object in front of our eyes. But, due to our scientific knowledge of cognitive neural mechanisms and their mathematical models, we know that these emotional neural signals are there. And it is easy to prove experimentally. As soon as the perception and understanding of the surrounding world does not work, we feel disharmonious, disturbed, or even threatened—this is a routine material for movie thrillers, which present situations that do not fit our concept-models. At the level of simple objects, this perception mechanism is mostly autonomous, like the workings of our stomach. As long as the stomach works perfectly, we do not notice its existence emotionally. But as soon as it fails, we feel it emotionally right away.

THE HIERARCHY OF THE MIND AND THE EMOTIONS OF THE BEAUTIFUL

The mind organization is an approximate hierarchy from objects to abstract ideas (Fig. 2.1). At every level of the hierarchy, top-down signals generated by concept-models are matched to bottom-up signals coming from concept-models recognized and understood at lower levels. Of course understanding involves aesthetic emotions at every level of the hierarchy. The mind hierarchy involves multiple levels of concept-models, from simple perceptual elements (like edges, or moving dots) to concept-models of objects, to complex scenes, and up the hierarchy... toward the highest concept-models. The highest concept-models near the top of the hierarchy are essential for abilities of the beautiful and spiritually sublime.

To understand this, let us first attend to the perception–cognition of a simple situation-scene, say, an office of a professor. It is not sufficient for our knowledge instinct to understand individual objects in the office such as books, shelves, chairs, the desk, and computer... we can sit in a chair or read a book, but this understanding will only take us so far (animals also understand objects). The knowledge instinct drives us

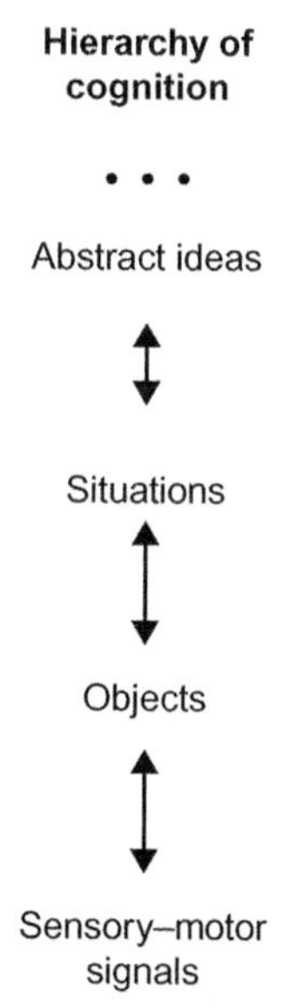

Figure 2.1 The hierarchy of cognition. Cognition is organized into an approximate hierarchy from objects to abstract concepts and higher up to the highest cognitive concepts of the meaning and purpose, closely related to the emotions of the beautiful.

to understand the concept "office" in its unity of constituent objects. A mathematical model of this process was developed by Perlovsky and Ilin. For understanding higher level abstract concepts, we have corresponding concept-models, such as "office." Similarly, we understand a "concert hall," and any other situation by using appropriate-level concepts that we have for this purpose. Let me repeat this word: purpose; every higher level concept and its mechanisms evolved in genetic and cultural evolution and in individual learning with a purpose to make unified sense out of many lower level concepts. In this process lower level concepts acquire higher level "sense," or a meaning of making up something "bigger," something more meaningful than their lower level meanings. In this way our understanding of the world can move from a "book" to an "office," to a "university," to an "educational system," and so on to concepts near the top of our minds. These "top" concepts "attempt" to make sense, to understand the meaning of our entire experience. We understand—perceive—feel them as related to the meaning and purpose of our lives.

Let us clarify this point. First, let me reiterate that even a simple object, when imagined with closed eyes is vaguer and less conscious than when perceived with opened eyes. But abstract concepts at higher levels of the mind hierarchy cannot be "perceived with opened eyes." Correspondingly, they are forever vaguer and less accessible to our consciousness than simple objects. The reason abstract concepts may seem crisp, clear, and conscious will be addressed in Chapter 3, Language and Wholeness of Psyche. Second, vaguer and less conscious concepts may also be mixed up with emotional contents. For example, talking about your favorite political party may require special effort to separate your conceptual understanding from your emotional involvement. In addition, we discussed that the process of understanding is motivated by aesthetic emotions, which therefore must be present at every level of the hierarchy. This is why concepts at the top of our mind are likely to be at once emotionally charged and less conscious. Many of my friends (scientists) when asked: "Does your life have meaning and purpose?" will reply with great doubts. Nevertheless, as soon as the question is reformulated: "So your life does not have any more meaning and purpose than that rock on the side of the road?" At this point most people agree that the idea of the meaning and purpose of life might be vague and barely conscious, but it is so important that we cannot live without it.

Life does not convince us that our lives have meaning and purpose; random deaths and destructions abound. But believing in one's purpose is tremendously important for survival; it is necessary for concentrating the will and power on achieving higher goals in life. This is why even a partial understanding of the contents of the highest concept-models is so important. When we feel that indeed our lives have meaning, in these rare fleeting moments we feel the knowledge instinct satisfaction at the highest level as an aesthetic emotion of the beautiful.

Similar arguments were made by Aristotle and Kant. Aristotle wrote that the beautiful is a "unity in manifold." The only way to understand the world in its unity, he wrote, is to understand it as if it had a purpose. Kant understood the beautiful as "purposiveness without purpose" or "aimless purposiveness" of the faculty of judgment; Kantian judgment corresponds to mechanisms of aesthetic emotions as discussed in the previous section and mathematically modeled by Perlovsky. In addition, "aimless" in Kant means that it is not aimed at satisfying lower bodily needs. Kant did not appreciate the need for the adaptation of concept-models and could not formulate the idea of the knowledge instinct. This caused him great difficulty; he gets around "aimless purposiveness" by emphasizing that it is not aimless, that it is highly spiritual, but without the idea of the knowledge instinct he could not provide a positive definition of the beautiful. The formulations discussed here might be clearer, because they are based on the mathematical theory.

The considered theory of the mind predicts that emotions of the beautiful are related to the meaning and purpose of life (and not necessarily to specific colors or patterns on a canvas, or specific sequence of musical keys). This is a revolutionary scientific idea about the beautiful and the meaning, which is radically different from anything taught in university courses on art. It also makes a revolutionary breakthrough in understanding the thoughts of Aristotle and Kant, making clear and simple the topics of many convoluted doctoral dissertations and "guides" to Kant.

Such a breakthrough in the understanding of most important issues considered by great minds over millennia requires some confirmations stronger than "trust me." Scientists know that the very first test of a scientific theory is its elegance and beauty; this includes Einstein, Poincare, Dirac. Still the best and final proof of a scientific theory is an experimental confirmation of its predictions. The theory discussed above made many predictions and some of them have been experimentally proven. Experimental confirmations of predictions about the nature of the highest meaning, emotions of the beautiful, and relations between them have been recently obtained, they are discussed in Chapter 6, Experimental Tests of the Theory: Beauty and Meaning.

CHAPTER 3

Language and Wholeness of Psyche

Contents

Abstract

The dual hierarchy of language and cognition is glued by unconscious emotions of prosody. This mechanism is fundamental for understanding cognitive functions of music. Language differentiates consciousness, but our mind also requires synthesis, the feeling of being whole.

THE DUAL HIERARCHY OF COGNITION AND LANGUAGE

The mind hierarchy as discussed earlier tacitly assumes a single hierarchy of cognitive models. To get closer to understanding musical emotions, we now consider interaction of language and cognition, the dual hierarchy of cognition and language; following is a summary of this theory, which is discussed in detail in publications by Perlovsky and coauthors.

The recognition that language and cognition are not the same, that these abilities are served by different mechanisms of the mind, began a revolution in 20th century linguistics initiated by Chomsky. Many psycholinguists and evolutionary linguists today disagree with Chomsky's complete separation of language from cognition and the denial of the evolutionary origin of language. Detailed discussions can be found in the given Literature. Here I summarize only the conclusions that are important for understanding the function of music.

What is the difference between language and cognition? Language is so important for thinking that it is difficult to comprehend what cognition would be without language. Do we think with words? Or do we use words only to label thoughts after they have been formulated in mind? How does cognition interact with language? Children typically acquire language by about the age of 5, by 7 they can talk about much of the

Music, Passion, and Cognitive Function.
DOI: http://dx.doi.org/10.1016/B978-0-12-809461-7.00003-0

content of culture. But they cannot act as adults. What exactly is missing in terms of neural mechanisms? How do children learn which words and sentences correspond to which objects and situations? Some people master language very well, while inept in the real world or when interacting with other people; contrary examples also abound. So, what are the mechanisms that make language and cognition so interdependent, and at the same time so separate? And what exactly are animals missing that they cannot learn language?

The main mechanism of interaction between cognition and language, according to the given Literature, can be modeled by the dual hierarchy (Fig. 3.1). In this model each concept-representation has two parts, the language part (a representation of a word or phrase) and the cognitive part (a representation of an object or event). When a child is born, these are vague neural placeholders that later acquire concrete content. By the age of five or seven, most of the language representations are crisp, clear, and conscious, but many of the corresponding cognitive representations may remain vague and unconscious. The reason is that individual learning of language relies on the surrounding language, in which contents of language representations, words, phrases, are "ready-made" for learning. By the age of four, everyone knows, e.g., about good guys and bad guys, but who can claim at 30 or 40 or 70 years, that he or she can use these concepts in real life error-free? Philosophers have argued about the meaning of good and evil for millennia. Even for everyday concepts, the linguistic parts are crisp and conscious in every child's mind, but it will take the rest of the child's life to acquire equally crisp and conscious cognitive representation. Representations of objects are acquired early, alongside with language, because we see objects ready-made for cognitive learning. But contents of abstract concepts do not exist in the world "ready-made." Not every combination of events is worth learning as a separate abstract concept, understanding of abstract concepts requires experience guided by language. It is likely that most cognitive concept-representations never attain equally conscious and crisp states to those of language. If a concept does not exist in language, it is likely that it does not exist in cognition, and corresponding events are not even noticed. This is why many people speak words without being fully conscious about what they say. These properties of language—cognition interaction are explained by the mechanism of the dual models. Language representation refer to facts of language, and not directly to events in the world. Cognitive representations combine language with experience and refer to events in the world.

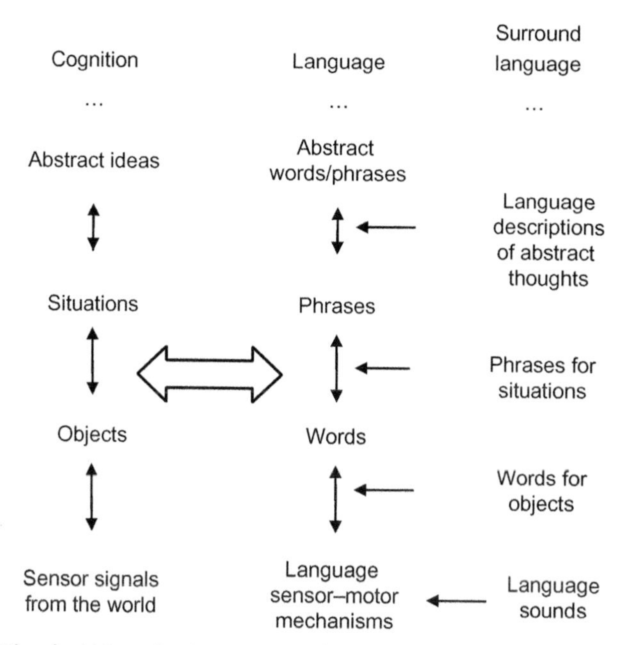

Figure 3.1 The dual hierarchy. Language and cognition are organized into approximate dual hierarchy. Learning language is grounded in the surrounding language throughout the hierarchy. Cognitive hierarchy is grounded in experience only at the very "bottom"; the cognitive hierarchy is constructed from experience guided by language.

A recent publication seems to support the dual model hypothesis. Franklin and coauthors have demonstrated that certain cognitions based in the right brain hemisphere in prelinguistic infants are rewired to the left hemisphere as language is acquired. Brain modules and neural connections involved in the dual models and knowledge instinct (KI) were discussed by Levine and Perlovsky.

The dual model has been fundamentally important for the emergence of the hierarchy of the mind. Learning should be grounded in experience. But concept-models of cognition are grounded in experience only at the lower levels of concrete objects; at this level human abilities are not much different from that of prehuman animals. Understanding situations and abstract concepts cannot be based on experience alone. The referenced publications discuss in detail why this is mathematically impossible: there are simply too many combinations of objects and events (more than all elementary events in the life of the Universe). A life's experience would never be sufficient to learn which combinations are meaningful to form abstract concepts.

To summarize, cognitive models at higher levels are learned based on both, life experience and language models. In this process language guides cognition: language identifies for cognition, which combinations of lower level concepts are meaningful for learning as a higher level concept. Language hierarchy is learned "ready-made" from the surrounding language at an early age. During the rest of an individual's life, the KI drives the mind to learn the cognitive hierarchy from life experience in correspondence with the language hierarchy. If a certain idea does not exist in a language, this idea does not exist in cognition and corresponding events would not even be noticed. Cognitive models are grounded in language.

DIFFERENTIATION AND SYNTHESIS

Learning within the dual hierarchy of the mind is driven by the KI, which operates with two main mechanisms, differentiation and synthesis. At every level of the hierarchy, it drives the mind to achieve detailed understanding by creating more specific, diverse, and detailed concepts—this is the mechanism of differentiation. At the same time, the KI drives us to understand various situations and abstract concepts as a unity of constituent notions. This mechanism of the KI, operating across hierarchical levels, creates higher meaning and purpose—this is the mechanism of synthesis. When operating in the hierarchy top-down, the KI drives differentiation, when operating in the hierarchy bottom-up KI drives synthesis.

The main "tool" of differentiation is language. Language gives our mind the culturally evolved means to differentiate reality in great detail. The evolution of language required the neural rewiring of circuits that controlled vocalization. The vocal tract muscles in prehuman animals are controlled from an old emotional center and voluntary control over vocalization is limited. Humans, in contrast, possess a remarkable degree of voluntary control over the voice, which is necessary for language. In addition to the old mostly involuntary control over the vocal tract, humans also have conscious voluntary control that originated in the cortex.

Correspondingly, conceptual and emotional systems (understanding and evaluation) in animals are less differentiated than in humans. The sound of an animal's cries engages its entire psyche, rather than concepts and emotions separately. A well-known example is the differentiated calls of vervet monkeys. The calls convey information about different types of predators nearby; however, understanding the situation (concept of danger), its evaluation (emotion of fear), and subsequent behavior (cry and jump on a tree) are not differentiated, each call is a part of a single concept-emotion-behavior-vocalization psychic state with very little differentiated voluntary control (if any).

Emotional evaluations in humans have separated from concept-representations and from behavior (e.g., when sitting around a table and discussing snakes, humans do not jump on the table uncontrollably in fear, every time "snakes" are mentioned). Differentiation of psychic states with a significant degree of voluntary control over each part gradually evolved along with language and brain rewiring.

Therefore language contributed not only to differentiation of our conceptual ability but also to differentiation of psychic functions of concepts, emotions, and behavior. This differentiation destroyed the primordial synthesis of the psyche. With the evolution of language, the human psyche began losing its synthesis, its wholeness. Although for prehuman animals every piece of "conceptual knowledge" is inextricably connected to the emotional evaluation of a situation, as well as to the appropriate behavior for satisfying instinctual needs, it is not so in humans. Most of the knowledge that exists in culture and expressed in language is not connected emotionally to human instinctual needs. This is tremendously advantageous for the development of conceptual culture for science and technology. Humans can engage in deliberate conversations, and if they disagree, they do not have to come to blows. But there is a heavy price that humans pay for this freedom of conceptual thinking: the human psyche is not automatically whole. Human knowledge accumulated through language is not automatically connected to instinctual needs; sometimes, culturally developed conceptual knowledge contradicts instinctual needs inherited from the primordial past. Moreover, various parts of knowledge may contradict each other.

Synthesis, the feeling of being whole, is closely related to the successful functioning of the highest representations at the top of the hierarchy of the mind, which unify experience and create the meaning and purpose of life. Therefore contradictions in the system of knowledge, the disconnects between knowledge and instincts, the lost synthesis may lead to internal crises and may cause clinical depression. When psychic states missing synthesis preoccupy a majority of the population, knowledge loses its value, including the knowledge and value of social organization—leading to cultural calamities, wars, and destruction. The evolution of culture requires a balance between differentiation and synthesis. Differentiation is the very essence of cultural evolution. But it may lead to an emotional disconnect between conceptual knowledge and instinctual needs, to lost feelings of meaning and purpose, including the purpose of any cultural knowledge, with a devastating impact on culture itself. Theoretical and experimental evidence suggest that different languages maintain different balances between the emotional and the conceptual.

CHAPTER 4
Music

Contents

Abstract

Evolution of consciousness and cultures requires both differentiation and synthesis, detailed understanding and wholeness. These fundamental mechanisms of the mind and culture require and oppose each other. Language leads toward fast accumulation of knowledge, which creates contradictions, cognitive dissonances leading to discarding knowledge. Evolution of cultures required cognitive mechanisms for overcoming this tendency to discard knowledge. This mechanism was strongly emotional voice evolving toward music. Music is necessary for the entire evolution of culture. While language evolves toward more conceptual and less emotional ability, music evolves toward more emotional and less conceptual ability. While language splits psyche, music restores its unity. We come to understanding why music has such power over us: we live in the ocean of grief created by cognitive dissonances, "in much wisdom is much grief"; and music helps us alleviate this pain.

DIFFERENTIATED KNOWLEDGE INSTINCT

In this chapter we discuss the main theory of this book: what constitutes the fundamental function of musical emotions in cognition, in the evolution of consciousness and culture, and why music affects us so strongly.

Differentiation and synthesis have to be balanced for the development of cultures and for the emergence of contemporary consciousness. Those of our ancestors who could develop differentiated consciousness could better understand the surrounding world and better plan their life. They had an evolutionary advantage, *if in addition* to differentiation they were able to maintain the unity of self required for concentrating the will. The evolutionary advantage of differentiated knowledge required balance between differentiation and synthesis. Here we examine the mechanisms by which music helps maintain this balance. The main idea of the theory discussed in this book is that maintaining this balance is the very

Music, Passion, and Cognitive Function.
DOI: http://dx.doi.org/10.1016/B978-0-12-809461-7.00004-2

fundamental function of music in cognition, and the reason for the evolution of this otherwise unexplainable ability.

History keeps a long record of advanced civilizations whose synthesis and ability to concentrate the will was undermined by differentiation. They were destroyed by less developed civilizations (barbarians) whose differentiation lagged behind, but whose synthesis and will was strong enough to overcome the great powers of their times. But I would prefer to concentrate on less prominent and more important events of everyday individual human survival—from our ancestors to our contemporaries. If differentiation undermined synthesis (the purpose of life and the will to survive), differentiated consciousness and culture would have never emerged.

Differentiation is the very essence of cultural evolution, but it threatens synthesis and may destroy the entire purpose of culture, as well as culture itself. This instability is entirely human, it does not threaten the animal kingdom because the pace of evolution and differentiation of knowledge from the amoeba to the primates was very slow, and synthesis was naturally maintained. This situation drastically changed with the origin of language; the accumulation of differentiated knowledge vastly exceeded the biological evolutionary capacity to maintain synthesis. Therefore, along with the origin of language another human ability must evolve, the ability to maintain synthesis.

Synthesis is violated by differentiation in the processes of cognitive dissonances. Cognitive dissonances are unpleasant emotions—feelings, discomforts due to contradictory knowledge. When new knowledge contradicts already existing knowledge, it is unpleasant. And it has been experimentally established that most often this discomfort is immediately resolved: the new contradictory knowledge is rejected fast and without reaching consciousness. I continue discussing cognitive dissonances in "Cognitive Dissonances, and Musical Emotions" section. Here I just emphasize that cognitive dissonances lead to rejecting new knowledge. Therefore to accumulate knowledge and continue cultural process, a new ability is needed, i.e., the ability to overcome cognitive dissonances created by language without discarding knowledge.

The dual language–cognition model considered in Chapter 3, Language and Wholeness of Psyche, suggests that this ability has to reside within language–cognition interaction; this ability has to generate strong emotions related to human voice, positive emotions to overpower the negative emotions of cognitive dissonances.

Thus a uniquely human ability evolved, the ability for music. The dual model in combination with the theory of cognitive dissonances results in a scientific theory that music evolved for maintaining the balance between differentiation and synthesis. After reviewing the arguments, we discuss the empirical and experimental tests verifying this theory.

Many scientists studying the evolution of language have come to the conclusion that originally language and music were one. In this original state the fused language-music did not threaten synthesis. Not unlike animal vocalizations, vocal sounds directly affected ancient emotional centers, connected the semantic contents of the vocalizations to instinctual needs and to behavior. This is how Jaynes explained the stability of the great kingdoms of Mesopotamia up to 4000 years ago. This synthesis was directly inherited from animal voicing mechanisms, and to this very day the voice affects us emotionally directly through ancient emotional brain centers.

I would like to emphasize that which have already being discussed: the fact that since its origin language has evolved to enhance our conceptual differentiation ability by separating it from ancient emotional and instinctual influences (here I mean "bodily" instincts, not instincts for knowledge and language). While language evolved in this more conceptual and less emotional direction, I suggest that "another part" of human vocalization ability evolved toward a less semantic and more emotional direction by enhancing the already existing mechanisms of the voice-emotion-instinct connection. As language was enhancing differentiation and destroying the primordial unity of psyche, music was reconnecting the differentiated psyche, restoring the meaning and purpose of knowledge, and making cultural evolution possible.

This was the origin and evolutionary direction of music. Its fundamental role in cultural evolution was that of maintaining synthesis in the face of increasing differentiation due to language. I would emphasize here that the number of cognitive dissonances is large: practically every word contains a new knowledge and therefore to some extent contradicts existing knowledge, otherwise a new word will not be needed. Correspondingly, a large number of musical emotions had to evolve to overcome cognitive dissonances. Although language evolved toward reduced emotionality, music evolved toward enhanced emotionality. Therefore theories of musical emotions, discussed in Chapter 1, Theories of Music, such as GEMS, which attempt to measure musical emotions in terms of emotional words, even if useful for some purposes cannot be successful for explaining musical emotions in principle. I now return to the basic mechanisms of the mind, including the knowledge instinct (KI), and analyze them in more detail in view of this theory.

Previous sections described the KI and the mathematical model of its mechanism, an internal mind "sensor" measuring the similarity between concept-models and the world, and the related mechanisms of maximizing this similarity. But clearly this is a great simplification. It is not sufficient for the human mind to maximize the average value of the similarity between all concept-representations and all experiences. Adequate functioning requires a constant resolution of contradictions not only between multiple, mutually contradicting concepts but also between individual concepts that are quickly being created within the culture and our slowly evolving primordial animal instincts. The human psyche is not as harmonious as the psyche of animals. Humans are contradictory beings; as Nietzsche put it, "human is a dissonance." Those of our ancestors who were able to acquire differentiated contradictory knowledge and still maintain the wholeness of psyche necessary for the concentration of the will and purposeful actions had a tremendous advantage for survival.

Therefore the KI itself became differentiated. It was directed not only at maximizing overall harmony but also at reconciling constantly evolving contradictions. This suggestion requires theoretical elaboration and experimental confirmation. As discussed, the emotions related to the KI are aesthetic emotions subjectively felt as harmony or disharmony. These emotions had to be differentiated along with the KI. Consider high-value concepts such as one's family, religion, or political preferences. These concepts "color" with emotional values many other concepts; and every contradictory conceptual relation requires a different emotion for reconciliation, a different dimension of an emotional space; Spinoza in the 17th century was the first to suggest that emotions referring to different objects are different emotions. In other words a high-value concept attaches aesthetic emotions to other concepts. In this way each concept acts as a separate part of KI: each concept evaluates other concepts for mutual consistency; this explains the notion of differentiated KI. Virtually every combination of concepts has some degree of contradictions. Otherwise, one concept, or even a simple instinctual drive would be sufficient for directing behavior.

COGNITIVE DISSONANCES AND MUSICAL EMOTIONS

Contradictions among knowledge, let us repeat, are called cognitive dissonances. These are negative emotions created by contradictions between pieces of knowledge—between conceptual representations. To illustrate these emotions consider an example: a young scholar receives two offers at once, one from Harvard and another from Oxford. Each offer alone would create strong positive emotions: satisfaction, pride, etc. These are well-understood basic emotions. But the choice between these two offers might be painful. This painful emotion is not related to bodily instincts, it is not a basic emotion; this is an emotion of cognitive dissonance, it is an aesthetic emotion. In this example the emotion can be very strong and conscious. Correspondingly, it would be resolved consciously, by weighing various aspects of these two alternatives. But the majority of cognitive dissonances are likely to be less conscious, or even unconscious. The barely noticeable, unpleasant emotions of the choice associated with knowledge can create a disincentive to knowledge and thinking. New knowledge creating cognitive dissonances often is quickly discarded. This indeed is well known and experimentally proven: the cognitive dissonance discomfort is usually resolved by devaluing and discarding a conflicting piece of knowledge. It is also known that awareness of cognitive dissonances is not necessary for actions to reduce the conflict, and these actions of discarding knowledge are often fast and momentary.

Let me repeat that cognitive dissonances often lead to discarding the contradicting knowledge. Everyone can observe it oneself. Watch carefully a usual conversation among people, not between a student and her Professor, but a normal conversation between regular people. Usually people do not listen to each other and immediately discard what they just have heard. We, scientists love to praise ourselves that we do not discard contradictions, that we enjoy contradictions because they give us food for thoughts, for creating theories overcoming contradictions. Yet what happens to discoveries that go against one's theory, or even simpler, against accepted theories. Well-known studies of the growth of knowledge established that new ideas are ignored, usually until the next generation of scientists. Great scientific discoveries may provoke not only fascination but also envy and rivalry. But worse, as established in the 20th century, the first reaction could be a cognitive dissonance, and as a result the novel is ignored.

The negative aspect of cognitive dissonance, discarding of knowledge, has received significant attention since Tversky and Kahneman were awarded the Nobel Prize in 2002. Cognitive dissonance is among "the most influential and extensively studied theories in social psychology." Still, the emotions of cognitive dissonances, their potential to destroy the drive for knowledge, and consequently the fundamental need to overcome their negative effects have not received sufficient attention. To overcome the negative effects of the emotions of cognitive dissonances, they must be brought into consciousness. This is the cognitive function of musical emotions. Music creates a huge number of differentiated emotions. Musical emotions help bringing to consciousness the emotions of cognitive dissonances, resolving them, and continuing the evolution of language, consciousness, and culture. The number of cognitive dissonances possibly is as large as the number of word combinations, practically infinite. Therefore aesthetic emotions that reconcile these contradictions are not just several feelings for which we can assign specific words. There is an almost uncountable infinity, virtually a continuum of aesthetic emotions. We feel this continuum of emotions (not just many separate emotions) when listening to music. We feel this continuum in Palestrina, Bach, Beethoven, Mozart, Chopin, Tchaikovsky, Shostakovich, The Beatles, and Lady Gaga ... (and certainly this mechanism is not limited to western cultures). It is proposed in this chapter that musical emotions have evolved for the synthesis of differentiated consciousness, for reconciling the contradictions that are entailed at every step of differentiation, and for creating a unity of differentiated Self.

The evolution of music, therefore, was necessary for evolution of culture. If not music, cognitive dissonances would have created disincentive to learning, including learning of language and learning any knowledge, which of course are the essence of culture. Musical emotions continue performing this function in cognition: overcoming cognitive dissonances so that culture continue evolving.

The origin of music resolving cognitive dissonances disappears from scientific sight at the dawn of human culture. Nikolsky wrote that Aurignacian culture (more than 30,000 years ago) developed conception of the Lunar calendars, "this re-oriented the entire lifestyle from local time to cosmic rhythms, which must have induced psychological stress on our predecessors who had to reconcile different notions of time, day/night, summer/winter, solar/lunar, as well as space." This "must have led to cognitive dissonances raising the need for compensating 'cognitive consonance' of music, and resulted in the substantial increase in harmonicity involving a transition from indefinite interval—based tonal organization to definite interval organization." This makes much sense in view of findings of Aurignacian pentatonic bone flutes.

Unification of musical modes into a single family in the great civilizations of the Bronze Age, also could be viewed within the context of growing cognitive dissonance. Rational harmonization of the entire compass of all available music tones appears as a natural progression of human culture. Inspired by correlative cosmologies, mathematically-based theories of music harmony catered to neurobiological need of the brain to reduce informational stress by employing a new strategy of organizing data and establishing ways for synthesis of new quality.

Here we come to understanding *why music has such a strong power over us*. Every piece of knowledge creates cognitive dissonances. We live in cognitive dissonances, in a sea of negative emotions created by them, "in much wisdom is much grief." And if not music we would continuously suffer negative emotions related to knowledge. Music helps us alleviate these negative emotions. Cognitive dissonances extend from minor everyday choices, such as a choice of drink between coca-cola and water, to life disappointments familiar to everyone, unrequited love, betrayal by friends, and loved ones. We do not notice negative emotions related to minor everyday choices, because we have a lot of emotions to overcome them. Strong dissonances related to disappointments with friends and loved ones are a major topic of popular songs. This is the reason we want to listen to Elvis Presley, The Beatles, and Lady Gaga. Most of popular songs help us to overcome these dissonances.

And of course there are ultimate emotional discomforts, dissonances related to our desire to live and at the same time the knowledge that our material existence is finite. Otherwise it is impossible to understand *why people enjoy sad music*. The most listened piece of music is Adagio for Strings by Barber, which is so sad it cannot be listened without tears. In 2004 listeners of the BBC's Today program voted Adagio for Strings the "saddest classical" work ever. In 2006 it was the highest selling classical piece on iTunes.

Music helps us to enjoy better our happy moments and to survive in the ocean of grief. This is why it holds such a sway over our souls.

Whereas language has differentiated the human Self into pieces, music has restored the unity of Self. This cognitive function of music is a scientific hypothesis, a theoretical prediction that has to be verified experimentally, and Chapter 5, Experimental Tests of the Theory: Music, discusses experiments confirming this theory.

CHAPTER 5

Experimental Tests of the Theory: Music

Contents

Abstract

The theory of musical emotions and their cognitive function has been tested in laboratory experiments. Theoretical predictions have been confirmed: music helps alleviate cognitive dissonances, accumulate knowledge, and unify the psyche. From choices of 4-year-old to judgments of seniors music helps cognition. Students taking musical classes outperform other students in all subjects. Is it due to music or due to inborn abilities?

The chapter discusses fundamental differences between hard and soft science. Why psychologists would claim no more than experiments "tentatively confirm" a theory, while physicists could use "confirm." A new area of science, physics of the mind transforms psychology toward a hard science. This will initiate a lot of rethinking.

THE FOX AND THE GRAPES WITH MUSIC

Let me repeat that holding contradictory cognitions creates cognitive dissonances, unpleasant uncomfortable emotions. It is known that this discomfort is often resolved by devaluing and discarding contradictory knowledge without much thinking. Cognitive dissonances could be useful adaptive reactions, promoting success and survival by enabling more efficient and consistent actions, or maladaptive, leading to irrational decisions and the devaluation of knowledge. Clearly, *it is essential to overcome cognitive dissonances,*

Music, Passion, and Cognitive Function.
DOI: http://dx.doi.org/10.1016/B978-0-12-809461-7.00005-4

which lead to the devaluation of knowledge. The hypothesis discussed in Chapter 4, Music, suggests that music helps performing such a cognitive function.

Recent experimental results tentatively confirm this hypothesis: it is possible that music helps us tolerate contradictory cognitions without devaluing contradictory knowledge. Masataka and Perlovsky have conducted experiments similar to classical studies of cognitive dissonances but with the difference of added background music. In the original experiment children devalued a toy if they were told that they couldn't play with it. This experiment has been reproduced hundreds of times with children and adults in various situations, confirming cognitive dissonance theory. The desire "to have" contradicts the inability "to attain." This contradiction creates a cognitive dissonance similar to the one described in Aesop fable, which is resolved in a similar way by discarding the contradiction: "if I cannot have it, it is no good and I do not need it."

However, when Masataka and Perlovsky reproduced the classical experiments with music playing in the background the toy was not devalued. In this study cognitive dissonance was experimentally created in 4-year-old children. An experimenter first elicited a ranking of toys. Later the experimenter suggested to a child not to play with the second-ranked toy and then left for a short time. According to the classical experiments, this would create cognitive dissonances and result in devaluing the second-ranked toy. This result was observed; the second-ranked toy became the last in rank. In another group of children the only difference was that background music was turned on when they were playing alone. In this group toy devaluation did not occur. The experiment was conducted twice to validate the finding. The statistical significance of the differences between the groups in the ranking of the "forbidden" toy with and without exposure to music was very high, $p < .001$ in each experiment. (This measure p answers the question: how probable is that the differences in valuation of the toy occur by random chance and not due to music? The answer is: such a probability is less than one in thousand in each of the two experiments; taking these experiments as independent, which is well justified, the "null hypothesis" that music has no effect has the joint probability of less than one in million). Therefore it was concluded that music can help to tolerate cognitive dissonances and hold in mind contradictory cognitions: one that the toy was attractive, another that the child did not play with it.

Additional details of the experiment *(copied with shortcuts from the above publication). Participants were 50 typically developing 4-year-old boys from several kindergartens in Kyoto and Aichi prefectures, Japan. They were randomly assigned to two groups 25 boys each (one group would be exposed to music during the experiment). We obtained written informed consent from the parents of all participants involved in our study. Prior to the experimental session, the experimenter spent several weeks at the kindergartens playing with the children, so that all of them could have known her well when the session started.*

The experimenter brought five different miniature cartoon monster figures (known as "Pockemon"). These figures were produced on the basis of images of monster characters that appeared in a TV cartoon, "Pocket Monster." All of the toys were extremely popular with children in Japan, particularly with young boys, and an opportunity to play with them was expected to be met with enthusiasm.

The experimenter led each participant into the experimental room, closed the door, and showed the participant the toys. She explained what each monster toy was, and allowed the participant to play with it briefly before moving on to the next one. After the participant became familiar with all the toys, the experimenter suggested a "question game" following which the participant was provided with an opportunity to play with the toys. The question game was aimed at eliciting a ranking of the toys from the most preferred toy (rank 1) to the least preferred (rank 5) toy. After the participant ranked the toys, the experimenter picked up the second-ranked toy, placed it on the table in the center of the room, arranged the remaining toys on the floor, and said: "I have to leave now for a few minutes to do an errand. But why don't you stay here and play with these toys while I am gone? I will be right back. You can play with this one [pointing], this one, and this one. But I don't want you to play with [mentioning the name of the second-ranked toy]. If you played with it, I would be annoyed. But you can play with all the others while I am gone, and I will be right back."

The experimenter then left the room. As she was leaving the room, she switched on an audio player, if the child in the room was in one of the two groups of the participants, exposed to music. The music continued to be played until the experimenter came back and switched off the player, whereas the children in the other group remained without such exposure to music. To summarize, the overall design of the experiment differed in only one respect between groups: music was played to one of the groups.

The experiment was conducted twice for improved certainty. A significance of a specific music piece played was not addressed - the important questions of the variety of musical emotions and their perceptions is discussed in later chapters. Mozart's sonata for two pianos in D major, K448 was played in the first experiment and Mozart's piano concerto No.23 in A major K488 was played in the second experiment.

The results of this experiment for the group without music are similar to a classical case of cognitive dissonances described in already discussed Aesop's fable "The Fox and the Grapes." When a fox sees high-hanging grapes that it cannot reach, the desire to eat and the inability to get the grapes creates cognitive dissonances in the fox's mind. The fox resolves cognitive dissonances by devaluing the grapes ("the grapes are sour"). Similar experiments were repeated many times and illustrated that it is a typical behavior of children and adults: "if I cannot have it then I do not need it." This way of resolving cognitive dissonances leads to discounting knowledge about what is good and important. The experiment of course also demonstrates the novel result that music helps resolve cognitive dissonances without devaluation of knowledge, that contradictory cognitions can be held simultaneously. Music helps to accumulate knowledge, to overcome contradictions, and to unify splits in consciousness.

DO STUDENTS SUFFER DURING TESTS AND COULD MUSIC HELP?

Another experiment by Cabanac, et al. reproduced the so-called Mozart effect: student's academic test performance improved after listening to Mozart. Bizarre claims appeared in mass media under this name: to have a genius child, pregnant women should listen to Mozart. Of course experts in music perception could not stand these ridiculous claims and many scientists set on proving these claims were wrong. Soon the Mozart effect was famously "debunked," any improvement was proved to be short-lived.

Despite these results, Cabanac, et al. used the Mozart effect to explore cognitive functions of music, specifically interactions of music with cognitive dissonances. The authors demonstrated first that students allocate *less* time to more difficult and stressful tests that create cognitive dissonances; this is expected from the cognitive dissonance theory. And second, they demonstrated that with music in the background students can tolerate stress, allocate *more* time to stressful tests, and improve grades.

Additional details of the experiment *(copied with shortcuts from Cabanac et al. (2013)).*

Method*. Sixty four participants in the experiment were students at De Rochebelle School (C.S.D.D), Quebec, Canada, 5th year high school, 14–15 years of age, both sexes. They answered a multiple choice type training test with 12 questions of their scientific course for fifteen minutes. After they had completed* *the test each received a form with 3 additional questions. Students were divided in two groups, 32 participants in each group, one group of 32 students listened to Mozart music during test, another group of 32 students listened to Koto music (Koto is a traditional Japanese music, similar to guitar). The two groups had identical grades performance. The theoretical prediction was that 'pleasant' music reduces cognitive dissonance stress inherent in tests and improves grade performance. In addition we wanted to study effects of 'unpleasant' music.*

The aim was to play two types of music: one calm and quiet to one group and, to the second group, a music widely different, vivid, and drawing attention; the music was without words. The nature and hedonicity of the environmental music played during the test had been selected by probing on other teenagers (one female and three males) who did not participate in the experiment. No explanation was given to that group beside the task of saying whether they liked or disliked different types of music they heard. The two melodies that received 4 votes were selected. Calm music was Mozart sonata in D for two pianos K.448, (used in Perlovsky and Masataka) especially the Andante; it was determined to be 'pleasant', and the other music was a koto solo with some disharmonious sequences, by Kuro Kami and Sakura Miyotote, determined to be 'unpleasant'.

It happened that some of the participants in the Mozart group, found that music unpleasant, and some from the koto group found it pleasant. Therefore the results were sorted, not on the account of the music heard, but on the pleasure or the displeasure experienced: 30 rated their music as pleasant, 21 as unpleasant, and 13 participants rated the music they heard as indifferent (zero hedonicity). These 13 participants served as a control and were labeled the 'unhedonic' group.

***Questionnaires**. Participants answered a multiple choice training type test with 12 questions of their scientific course for fifteen minutes. All questions concerned theoretical knowledge of the muscular and skeletal systems. At the end of their academic tests the participants answered two short questionnaires on separate pages, the Page A, probed their behavioral performance and the Page B, probed their experience. This protocol was arranged that way in order to avoid drawing the participants' attention to the environmental music and to their awareness aroused by the Page A questions. The music had been stopped at the end of the test, i.e. before these final questionnaires were opened.*

On Page A the participants were requested to:

write the exact time of their completion of the academic test; thus providing their individual duration

rate from 0 to 10 how difficult they had found the test

write from 0 to 100 the grade they expected to have earned

rate from 0 to 10, the intensity of their stress

On Page B the participants answered the following two questions:

have you been aware that music was played during the test? Answer Y/N.

did you like it? Rate your pleasure/displeasure experience, as a number between −5 and +5, with the following landmarks: −5 very unpleasant, −3 unpleasant, 0 indifferent, +3 agreeable, +5 very agreeable.

Results. *Grade performance and hedonicity of music. Grades earned by students in pleasant music condition were higher than for unpleasant or unhedonic. The differences are statistically significant. This tentatively confirms the first fundamental theoretical prediction.*

Then we evaluated the hypothesis that the hedonicity of music modulates the tolerance for cognitive dissonance. To isolate the effect of cognitive dissonance stress on reducing duration of tests, we compute regression of duration on two variables, difficulty and stress, and we could separately evaluate their effects and interactions with music. In this way we received the second fundamental result of the current report that deals with the cognitive function, origin, and evolutionary causes of music: music helps overcoming morbid consequences of cognitive dissonances. Since multiple choice tests require holding and evaluating contradictory cognitions, students are expected to experience cognitive dissonances resulting in a stress. Thinking, accumulating knowledge, and making choices involves CD, which causes stress. Thinking is stressful. This stress reduces time humans allocate to thinking. Stress reduces duration of tests. This effect is highly statistically significant. Whereas naively one could expect that more stressful tests should require more time, the results demonstrate that when the effect of difficulty is separated, the effect of stress (without music effect) is opposite from this naïve expectation. Stress reduces duration because stress is unpleasant and tolerating stress is difficult. If humans in their evolutionary development would not be able to overcome this morbid consequence of cognitive dissonances, human culture would not evolve to more knowledge and to ability for thinking.

Using 'pleasant' and 'unpleasant' music in this experiment was a first experimental step toward assessing the role of diverse musical emotionality in interaction with cognitive dissonances. No conclusive results were obtained, apparently this type of evaluation requires more specific musical emotions and cognitive conditions. As theoretically predicted in chapter 4, cognitive dissonances are expected to lead to a huge number of different emotions, and the correspondingly huge number of musical emotions are needed to overcome them. We are still far away from experimentally measuring a very large number of musical emotions. The next section describes another experimental step in this direction.

Experimental results reported in this section tentatively confirm fundamental theoretical predictions: *music evolved for helping to overcome a morbid consequence of cognitive dissonances, discarding contradictory knowledge. I call it morbid, because every knowledge contradicts to something already known, and therefore cognitive dissonances if not overcome would prevent knowledge accumulation, the essence of culture. Pleasant music helped keeping in mind contradictory cognitions in stressful thinking. With pleasant music students were able to tolerate stress and devote more time to stressful thinking; this effect was highly statistically significant. This tentatively confirmed a theoretical prediction that pleasant music helps overcoming cognitive dissonances and leads to grade improvement.*

CONSONANT AND DISSONANT MUSIC VERSUS COGNITIVE INTERFERENCE

I repeat a theoretical prediction from Chapter 4, Music, there are thousands of different musical emotions. Experimental studies of musical emotions are still far away from demonstrating that. The current section discusses a step in this direction: experimental results on studying musical consonances and dissonances in their interaction with cognition; more specifically—cognitive interference, phenomena described below, closely related to cognitive dissonances. In addition to previous results about the function of music in helping to overcome cognitive dissonances, this section differentiates the roles of consonant and dissonant music.

Debates on the origins of consonance and dissonance in music have a long history. Some scientists argue that consonance judgments are culture-related. Others favor a biological explanation for the observed preference for consonance, possibly related to the Helmholtz theory discussed in Chapter 1, Theories of Music. Here I follow Masataka and Perlovsky (2013) and describe experimental confirmation that this preference plays an adaptive role in human cognition: it reduces cognitive interference (discussed below). The experimental results reveal that exposure to a Mozart minuet mitigates interference, whereas, conversely, when the music is modified to consist of mostly dissonant intervals the interference effect is intensified.

To create cognitive interference, experiments in this publication used a prototypical "Stroop interference task." In the task, typically, a color word such as GREEN appears in an ink color such as red. If the participant's task is to read the word and ignore the color (e.g., say "green"), there is no difficulty reading the word compared to reading it when printed in standard black ink. However, if the participant's task is to name the ink color and ignore the word (e.g. say "red"), there is considerable difficulty. Reading the word interferes with naming the color. This is the phenomenon of Stroop interference.

Two specific hypotheses were tested in this experiment. The first hypothesis has been that music with more consonant intervals would reduce cognitive interference relative to that when the same person was tested without exposure to any music. The second hypothesis has been that music with more dissonant intervals would increase cognitive interference relative to that when the same person was tested without exposure to any music.

> ***Method***. *Participants in the experiment were 25 typically developing healthy children aged 8–9 years old and 25 healthy elderly adults aged 65–75 years old. They were asked to name the ink of a color word that designated a color incongruent with that of the ink of the word ('Incongruent test session'). Also, the same participants were tested in a 'Neutral testing session', in which they were asked to name the ink of a color of a non-word string of letters, i.e., XXX. Both sessions were repeated under three conditions: (1) with exposure to music containing predominantly consonant intervals (Consonant condition), (2) with exposure to music containing predominantly dissonant intervals (Dissonant condition), and (3) without exposure to any music (Control condition). In every session and condition, the performance of the participants was measured as reaction time (RT) to response and error rate (ER) of the response.*
>
> *The consonant music used was the original version of one of Mozart's minuets, most of which consisted of consonant intervals. For dissonant music we used a modified version of the minuet, most of which consisted of dissonant intervals.*
>
> ***Results***. *All hypotheses have been tentatively confirmed with high statistical significance. RTs were significantly longer for Incongruent session vs. Neutral session ($Ps < 0.001$). RTs for Incongruent sessions were significantly shorter under the Consonant condition than under the Dissonant or Control conditions ($Ps < 0.001$). Similarly, the RT under the Dissonant condition was significantly longer than that under the Control condition ($P = 0.032$). In Neutral sessions RTs did not differ among the three conditions ($Ps > 0.711$). The Stroop effect results in cognitive interference as expected, consonant music helps to overcome cognitive interference, and dissonant music increases the interference.*
>
> *The results obtained using ER in the children were similar to the results obtained using RT, confirming the effects discussed above. The results of the experiment with the elderly adults were strikingly similar to those with the children. All of these results gave additional confirmation of the hypotheses.*

The conclusion is that consonant music may have an important cognitive function: help overcoming cognitive interference. This gives further support for the hypothesis about the fundamental cognitive function of music: it helps to resolve cognitive interference, cognitive dissonances and facilitates human evolution.

Another issue is the role of consonant versus dissonant music and their relations to pleasure of music. It is known that infants and even newborns exhibit strong perceptual preferences for the original minuet containing mostly consonant intervals over its modified dissonant version. Results reported in this section support recent findings discussed above that the effect of music on cognitive dissonances depends on the hedonicity of music: pleasant music better helps to overcome cognitive dissonance than unpleasant music.

Still this is just a first step toward evaluating interactions between multiplicity of musical emotions and cognitive dissonances. Drawing conclusions about the connection between musical consonance and hedonicity requires caution since dissonant and sad music could also be sources of pleasure as discussed in Chapter 4, Music. Because of fundamental importance of Chapter 4, Music, predictions, here we repeat some of these discussions. Now we will discuss possible cognitive functions of musical dissonance, an unresolved issue in psychology and musicology. We demonstrated that the modified version of the minuet containing more musical dissonances increased the cognitive interference.

This result, let us repeat, might be compared to the findings of a study by Thompson, Schellenberg, and Husain (2001) whose aim was to debunk the popular version of the "Mozart effect" and to demonstrate that any improvement on cognitive tests after listening to Mozart is not specific to music. Those authors reported that whereas Mozart's music results in some improvements of cognitive test scores, Adagio by Albinoni (sad, slow music; the actual author might be R. Giazotto) results in lower scores on the same cognitive tests; they demonstrated that mood and arousal may account for the "Mozart effect." The authors of the study however did not investigate a fundamental question of why the Albinoni adagio is among the most popular pieces of music.

So what could be the cognitive function of musical dissonance and of music per se and evolutionary reasons for music evolution? Our analysis of the music cognitive function suggested that it is to overcome a large number of cognitive dissonances between virtually any two cognitions. These include stress that arises in many complicated and difficult life conditions much more difficult and trying than those evoked by the Stroop effect and more important than trivial cognitive improvement studied by Thompson, Schellenberg, and Husain (2001).

A specific aspect of this question is why sad music is pleasurable. One of the most popular pieces of western classical music is Adagio by Barber, which is sad, slow, and highly dissonant, like Adagio by Albinoni. According to our hypothesis sad music helps to overcome dissonance arising from difficult life conditions, including the ultimate deaths of close people and oneself (the dissonance between the feeling of the infinity of the spirit and the knowledge of death). In general, any two cognitions involve a cognitive dissonance. Possibly the cognitive dissonance between any two cognitions involves its own shade of emotion, and overcoming each cognitive dissonance requires a special musical emotion. This hypothesis implies a potentially large number of cognitive dissonances, interferences, and a correspondingly large number of musical emotions. Music evolved for helping to overcome the predicament of stress that arises from holding contradictory cognitions, so that knowledge is not discarded, but rather can be accumulated, and human culture can evolve.

Our experimental results emphasize a need for further research studying multiple emotions and for determining the dimensionality of these emotional spaces. This problem has not been solved, and the current chapter reports steps in this direction. The consonance–dissonance dimension explored here is related to hedonicity (pleasure or displeasure) perceived in music; however the potential pleasure from sad dissonant music makes this connection nontrivial. Possibly music is perceived as pleasant if it resolves cognitive dissonances and interferences important for a listener. Music pleasant for many people resolves dissonances and interferences important for many of us.

MUSIC AND ACADEMIC PERFORMANCE

Experiments conducted by Cabanac et al. (2013) demonstrated that students taking musical classes outperform other students in all subjects, whereas before musical classes grades were similar. This provides an additional confirmation of predictions of this book theory about the cognitive function of music. Music has an important cognitive function counterbalancing cognitive dissonances that lead to the devaluation of knowledge; music makes possible accumulating contradictory knowledge. This is essential for the entire human evolution and significant for understanding the cognitive function of music, its origin, and evolution—the issues remaining mysterious for 2500 years. Here we continue this line of inquiry by addressing a question if long-term study of music systematically improves academic performance.

Methods. *The population studied consisted of students from a secondary school of the province of Québec Canada. The students belonged to the* International Baccalaureat *program for which they were selected on their first year of secondary school based on their high grades in previous years. They formed a homogenous group in terms of their grades. During the first 2 yrs of their secondary school curriculum (levels 1 and 2), music was compulsory with two courses taking nine days per period. Over the third, fourth, and fifth years (years 3, 4, and 5), music courses were optional and students had to choose one option between plastic art (painting and sculpture), dramatic art, and music. During the first two years the student academic performance was similar in this highly selective and performing population. Over the following three years all students were still of similar academic standard, but any student who disliked music was free to choose another optional course, and all the students who liked music could continue to take it at the school. A new experienced teacher, very skilled and much liked by students, had been teaching over the last three years of the music course. Several of his students were eventually able to enter the* Conservatoire de music *after these three years.*

The mean grades for the academic year 2011–2012 were recorded for three different school years, corresponding to the third year (n = 196 students), the fourth (n = 184 students) and the fifth (which is the senior class of the secondary school; n = 180 students). Students were of both sexes, aged 14–15, 15–16, and 16–17 years old (for the corresponding school year). It is important to keep in mind that all the students were among the top grade level of their school, whether they selected music courses or not.

The three different school years were analyzed as three separate groups. From all the test results available we selected only courses with quantifiable performance including: sport, science, mathematics, French, English, history, chemistry, physics, Spanish, ethics, present-day world.

Results. *Figures 5.1, 5.2 and 5.3 illustrate striking results. Each year, the mean grades of the students that had chosen a music course in their curriculum were higher than those of the students that had not chosen music as an optional course. This tendency is true regardless of the topic of the course*[1].

[1] There are two exceptions to this statement. Of the 25 courses rated over 3 years (Figures 5.1, 5.2 and 5.3) there are only 2 cases when non-musical student scores are higher than musical student scores (History year 2, French year 3); in both cases the differences are of low statistical significance.

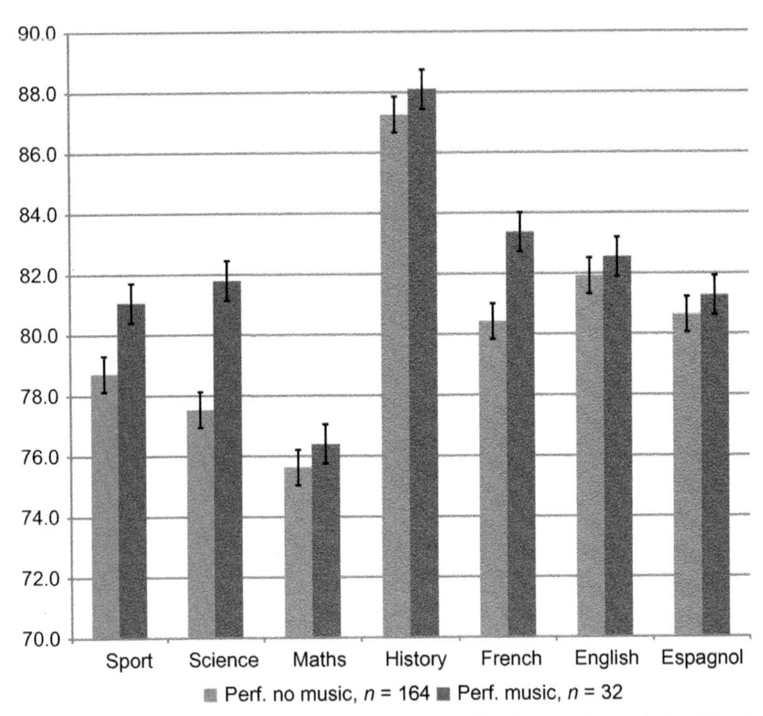

Figure 5.1 The mean grades for different courses for the students (14–15 yrs) from the third school-year of secondary school, during the year 2011–12.

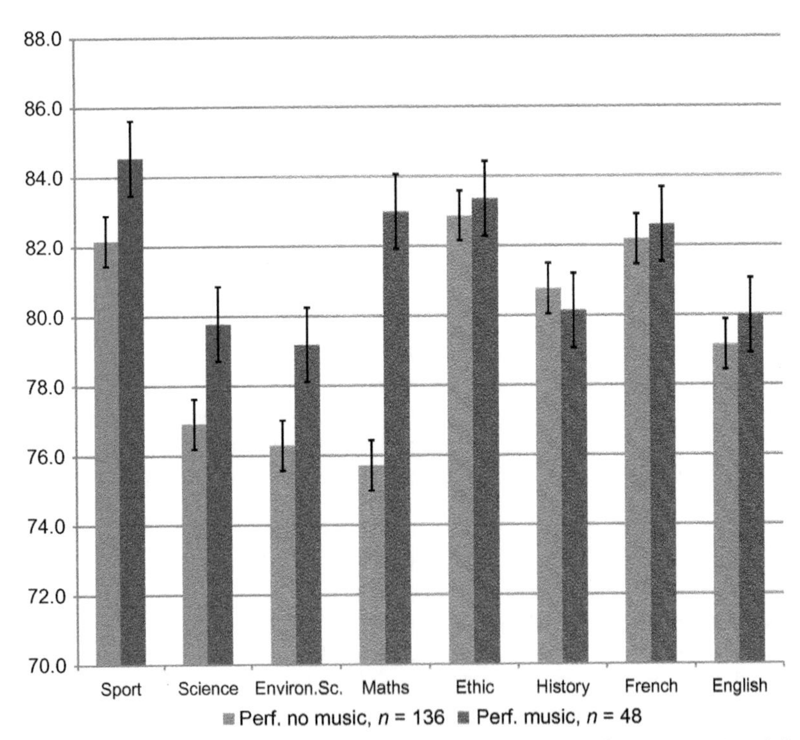

Figure 5.2 The mean grades for different courses for the students (15–16 yrs) from the fourth school-year of secondary school, during the year 2011–12.

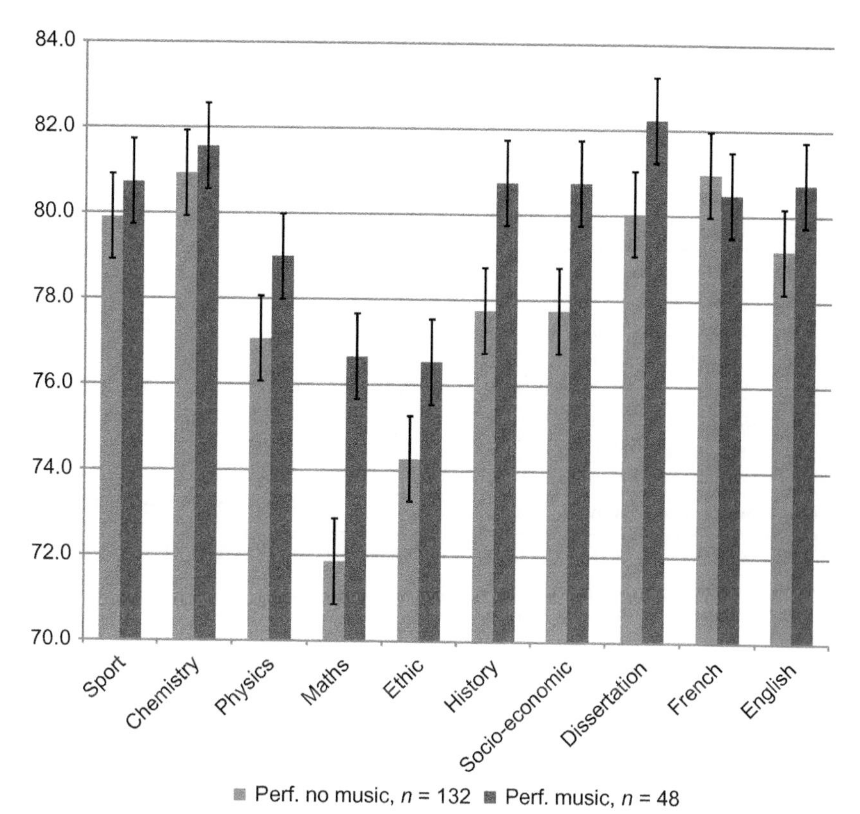

Figure 5.3 The mean grades for different courses for the students (16–17 yrs) form the fifth school-year of secondary school, during the year 2011–12.

Statistical significance of these results is very high. The probability of null hypothesis (H0: music does not positively affect academic performance), according to T-test over all data is very low, $p < 0.001$. For individual subjects over years statistical significance varies, with many cases reaching high statistical significance, as shown in the following Tables 1, 2, 3 (corresponding to Figures 5.1, 5.2, 5.3).

Table 1 Statistical significance of the results in Fig. 1: T-test probabilities of the null hypotheses, H0, are low for Science and French (H0: music has no positive influence on academic performance as measured by the mean grade for each course for the students (14–15 yrs) from the third school-year of secondary school, during the year 2011–12)

subject	Sport	Science	Math	History	French	English	Spanish
$p<$	0.11	0.00	0.43	0.32	0.06	0.31	0.34

Table 2 Statistical significance of the results in Fig. 2: T-test probabilities of the null hypotheses, H0, are low for Sport, Science, Environmental science, Mathematics (H0: music has no positive influence on academic performance as measured by the mean grade for each course for the students (15–16 yrs) from the fourth school-year of secondary school, during the year 2011–12)

subject	Sport	Science	EnviSci	Math	Ethics	History	French	English
$p<$	0.00	0.01	0.02	0.00	0.35	0.33	0.33	0.25

Table 3 Statistical significance of the results in Fig. 3: T-test probabilities of the null hypotheses, H0, are low for Mathematics, History, Socio-economics, (H0: music has no positive influence on academic performance as measured by the mean grade for each course for the students (16–17 yrs) form the fifth school-year of secondary school, during the year 2011–12)

subject	Sport	Scienc.	EnvSci	Math	Ethics	History	Socio-econ.	Diss.	French	English
$p<$	0.26	0.12	0.15	0.03	0.12	0.03	0.03	0.35	0.11	0.13

These results confirm that music has a link to cognition. They do not indicate causality: whether the students have better grades because they practice music from time to time or whether they chose music because they are better at school. However these data correspond to the previous results: music helps students to hold contradictory knowledge despite stress from cognitive dissonance caused by this contradictory knowledge. We wanted to understand why students quit music courses and conducted a survey of students who had quit the music course. 15 students responded to this survey (some indicated more than 1 reason): dislike music (2 answers), curiosity to try something else (7 answers), like music but do not like playing with an instrument (6 answers), already practicing music outside of school on a regular basis (3 answers).

DISCUSSION

This study confirmed a long line of research linking music and cognition. Students taking musical courses achieve better performance on all other subjects. When combined with results in previous sections, this gives a strong support to the hypothesis that music helps overcoming stress due to cognitive dissonance, helps accumulating knowledge, and music is fundamental for human evolution.

Existing experiments have not directly addressed the issue of causality: does music training causes improvements in cognition or high-functioning children are more likely than other children to take music lessons, and that they also differ in personality? The issue of causality has not been resolved in that publication. Possibly, combining a theoretical prediction of musical emotions being the mechanism resolving cognitive dissonance, experimental confirmations of this hypothesis, and evidence for correlation between engagement with music and other cognitive achievements in this section give solid evidence for causality of interaction between music and cognition.

CONCLUSION

The current section contributes to establishing relations between music education and academic performance. We have demonstrated that students selecting musical courses perform better than those declining such courses, despite equally high initial achievements. This further contributes to accumulating theoretical and experimental evidence that music helps overcoming stress caused by cognitive dissonance, and helps accumulating knowledge, which is fundamental to human evolution. In addition to addressing this fundamental scientific question about the origin and evolution of music, it also contributes to the ongoing debate about needs and usefulness of musical education.

CAN EXPERIMENTS CONFIRM A THEORY? PHYSICS OF THE MIND

In conclusion of this chapter devoted to experimental tests of the theory, let me make a comment, can we expect that an experiment confirms or tentatively confirms a theory. In physics, if an experimental observation corresponds to the theoretical prediction it is usually understood as a confirmation of the theory. For example, Newton's theory of gravitation was used to predict the existence of a planet Neptune. When a telescope was pointed to a predicted direction, indeed a new planet was discovered. This was considered a confirmation of the theory. Einsteinian theory of relativity predicted that light from stars should bend near sun. When confirmed experimentally, this was considered a confirmation of the theory. In psychology more cautious statements are made: an experiment, at best, may *tentatively* confirm a theory. In addition, psychologists usually require probabilistic statements of the sort made throughout this chapter that a *random* chance that an observation is confirmatory (null hypothesis) is highly improbable, say the random chance $p < .05$.

Are psychologists more exact in their scientific conclusions? Or is there a fundamental difference between psychology and physics, which warrants stronger statements by physicists? Let me suggest that indeed, historically, there have been a fundamental difference between psychology and physics in this regard. Well-known philosophers of science, including Popper, Kuhn, Meehl, attempted to formulate this difference between physics and psychology, between "hard" and "soft" sciences. I will summarize now most important conclusions and I will comment on the coming fundamental changes in this regard. Psychological theories and experiments historically addressed narrow fields using uncertain theories (e.g., matching a linear equation to a set of experimental data points), where theoretical predictions *could randomly* match experimental measurements. Physical theories and experiments address universal fields, where a random match of observations to theoretical predictions is impossible (such as prediction of a new planet). A fundamental difference between physics and psychology is in that physical theories are based on few fundamental principles, such as Newton's laws, from these few principles a theory is derived correctly predicting a wide area of observations, such as movements of all planets, comets, and other celestial bodies in the solar system, as well as falling of an apple from a tree. In psychology similarly wide scope theories based on few fundamental principles have not been possible until recently and developing such theories have not been even a goal of research. This changed recently due to neural imaging research and better understanding of the working of the mind. The mind *fundamental principles* are being identified; some of them have been summarized in Chapter 2, Mechanisms of the Mind: From Instincts to Beauty.

Based on these fundamental principles of the working of the mind, a wide scope mathematical theory has been derived, which is the basis of this book. Although this book avoids using mathematics, mathematics is fundamental for deriving a valid theory discussed in Chapter 2, Mechanisms of the Mind: From Instincts to Beauty. Discussion in this book is limited to conceptual-logical arguments; this is OK when the correct conclusions are known to be supported by a mathematical theory. But in absence of a mathematical theory conceptual-logical arguments could lead in a wrong direction. This is true, e.g., about psychology; logical arguments about working of the mind systematically lead psychologists in a wrong direction because the mind is not logical and this has not been understood. This is true even about distinguished scientists, such as Dawkins, Pinker, and others.

The theory in this book explained mechanisms of the mind, including perception, cognition, and interaction of cognition and language, mechanisms that were not understood for decades and millennia. Attempts to develop artificial intelligence based on mechanisms of the mind continued since the 1950s. Although artificial intelligence used mathematical methods, it did not follow the fundamental methodology of physics: (1) elucidating basic principles; (2) mathematical theory derived from these principles; and (3) predictions of this theory that could be verified experimentally. Artificial intelligence has been developed by mathematicians, and mathematics is fundamentally different from physics; whereas physics uses mathematics for describing the world, mathematics does not. That is the reason that mathematical attempts to model the mind have not been successful until development of the current theory. Based on the current theory and related understanding of the mind mechanisms, a number of artificial intelligence problems have been solved. Such a wide scope theory encompassing psychology and artificial intelligence, based on few fundamental principles, makes psychology a "hard" science similar to physics. This new science is called physics of the mind, it is a part of currently developed Physics of Life, and there is a highly ranked journal devoted to this new science.

Physics of the mind describes a wide field, predictions of this theory, including relationships between emotions of the beautiful and "higher" meanings are discussed in Chapter 2, Mechanisms of the Mind: From Instincts to Beauty, the role of musical emotions in cognition and evolution of culture are discussed in Chapter 4, Music, and the rest of the book. In this chapter the book discusses psychological experiments evaluating validity of the theory predictions. According to terminology standard in psychology, these experiments tentatively confirm the theory with appropriately high probabilistic measures. These probabilistic measures, as customary in psychology, account for a small possibility that the theoretical predictions are matched by experimental observations by a random chance in a single experiment. But there is no probabilistic technique that can account for certainty or uncertainty of theoretical explanations of problems that could not have been explained for decades or millennia, such as the role of music in cognition and evolution. Such a wide scope theory predicting experimental observations in wide fields should be evaluated similarly to theories in physics and considered experimentally confirmed.

CHAPTER 6

Experimental Tests of the Theory: Beauty and Meaning

Contents

Abstract

Aesthetic chills, strong aesthetic emotions have been used for experimental tests of theoretical predictions about the nature of the beautiful and its relations to the highest meaning and purpose. What is the beautiful, what are the highest meaning and purpose? Theoretical predictions are confirmed: the beautiful is defined not by shapes, or forms, or colors, or story tensions, all of these are important, still the beautiful is what makes human life meaningful and purposeful. Emotions of the beautiful are related to understanding contents of mental models near the top of the mental hierarchy. Kantian aesthetics is made simple. Courses on aesthetics will have to change their contents.

AESTHETIC CHILLS

In Chapter 2, Mechanisms of the Mind: From Instincts to Beauty, I discussed the hierarchy of human cognition from instincts at the "bottom" to the beauty at the "top." Chapter 2, Mechanisms of the Mind: From Instincts to Beauty, resulted in a stunning prediction that at the "top" of the human mind hierarchy there are emotions of the beautiful and representations of the meaning and purpose; contents of these representations are inaccessible to consciousness. This prediction makes a revolution in aesthetics: emotions of the beautiful are not related to sex, as many suspect, not related to forms of objects of art, they are related to understanding the meaning and purpose

Music, Passion, and Cognitive Function.
DOI: http://dx.doi.org/10.1016/B978-0-12-809461-7.00006-6

of life. Is it possible to verify experimentally this theoretical prediction, which seems highly unlikely?

Relating the beauty to meaning is remarkably self-consistent theoretically. And it follows in a great philosophical tradition, at least since Kant. But is it possible to prove it experimentally? I thought it will take many years before it would be attempted, for the theory discussed in Chapter 2, Mechanisms of the Mind: From Instincts to Beauty, aspires to *define* highly personal emotions and contents, *the beautiful, the meaning and purpose, and their relations*. Is it really possible to make a breakthrough progress in this area discussed by so many great thinkers since Aristotle or even before? Recently a colleague suggested that he knows how to do it. Our joint publication with this experimental verification has been printed, Schoeller and Perlovsky (2016), and this chapter summarizes its main ideas and results. In this chapter "we," "us" refer to authors of this publication.

This experimental study concentrates on special kinds of strong aesthetic emotions, aesthetic chills (chills unrelated to changes in temperature level). Nabokov described them in the following way: "It seems to me that a good formula to test the quality of a novel is, in the long run, a merging of the precision of poetry and the intuition of science. To bask in that magic a wise reader reads the book of genius not with his heart, not so much with his brain, but with his spine. It is there that occurs the telltale tingle even though we must keep a little aloof, a little detached when reading."

The results of this study suggest that aesthetic chills are positively correlated with curiosity (thus relate to knowledge and knowledge instinct [KI]), inhibited by exposing the participants to an incoherent prime prior to the chill-eliciting stimulation (thus relate to meaningful and purposeful experience; "additional details" below discuss a standard technique of "prime"), and that a meaningful prime makes the aesthetic experience more pleasurable than a neutral or an incoherent one (thus the pleasure of beauty is related to the meaning and purpose!). Aesthetic chills induced by narrative structures are related to the pinnacle of the story, have a significant calming effect, and participants in the experiment describe a strong empathy for the characters of the story. This chapter discusses the relation between meaning-making and aesthetic emotions at a psychological, physiological, narratological, and mathematical level and proposes a series of hypothesis to be tested in future. Most important, it tells us something new about emotions of the beautiful, about meaningful-purposeful experiences, and relates these emotions and experiences.

Aesthetic chills are universal emotions. This phenomenon invokes a wide variety of issues some of them discussed in the book, including the function of music in the cognitive system, the relation between cognition, social recognition and empathy, intelligence and collective intelligence, fear and expectations, aesthetic emotions and natural curiosity, aesthetic emotions, meaning and the KI, and, at a more general level, the function of artistic, scientific, and religious behavior in human societies.

To understand, describe, and predict aesthetic chills, one needs not only to be able to assess what can elicit them but also what can suppress them. Therefore this experimental study of beauty and meaning is not only concerned with how the related emotions might emerge but also by the conditions preventing this. Therefore the first task to understand the phenomenon of aesthetic chills is to determine a set of elicitors that provoke them and inhibitors that interrupt them. Seeking for logical simplicity and explanatory power, we conducted a series of experimentations related to the model of the KI discussed in Chapter 2, Mechanisms of the Mind: From Instincts to Beauty. First this chapter reviews the relevant data concerning aesthetic chills.

The processes of changing KI, its increase in the process of understanding, and its decrease when encountering new situations are continuously going on at multiple hierarchical levels. A simplified illustration of this process at a single level is shown in Fig. 6.1 (KI is shown in solid, its speed of change is aesthetic emotion in dashed). Understanding this figure and discussion of related mathematics is not necessary, use it to the extent it is helpful). The change in KI, as discussed, models aesthetic emotions, shown in red (dashed). At lower hierarchical levels, these

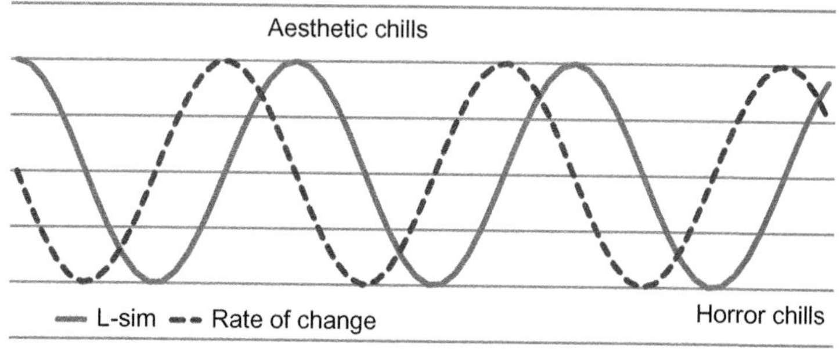

Figure 6.1 Notional illustration of KI changes. Satisfaction of KI, aesthetic chills, corresponds to max similarity between mental representations and external stimuli, L, and its derivative, $L' = 0$; dissatisfaction of KI, horror chills, corresponds to min L, and $L' = 0$. Vertical axis is similarity L and horizontal axis is time in arbitrary units.

emotions could be below a level of consciousness. Near the top of the hierarchy, they correspond to understanding of the highest meaning and may be experienced as emotions of the beautiful and aesthetic chills. The strongest emotions are experienced near the peak of the dashed curve in Fig. 6.1, which do not coincide with the peak of understanding, solid. At the moments of peaks of understanding changes slow down, the peak experience is a dynamic equilibrium. Later I compare this description to how participants in this experiment describe their peak experiences involving aesthetic chills.

When considering aesthetic chills, one notices that there seems to be two opposite types of chills: chills that are elicited by the subject's greatest fear and chills that are elicited by the subject's greatest hope. Chills are also known to be elicited by tonal music of both positive and negative valences. The theory in Chapter 2, Mechanisms of the Mind: From Instincts to Beauty, leads us to the hypothesis that chills might correspond to an event where KI similarity reaches a local maximum or minimum value (Fig. 6.1), and the rate of change of the similarity function tends toward a null value during chill. Aesthetic chills may be elicited in artistic, scientific, and religious contexts. They have mainly been studied in the laboratory using tonal music as an elicitor.

Musical chills are known to provoke changes in autonomic and other psychophysiological activity. Measurements of regional cerebral blood flow identified changes in structures associated with the brain reward circuitry. Analysis revealed increases in left ventral striatum, dorsomedial midbrain, and decreases in right amygdala, left hippocampus/amygdala, and ventral medial prefrontal cortex. The pattern of activity observed in correlation with music-induced chills is *similar to that observed in other studies of highly pleasurable emotions* and a neural theory of how pleasure related to knowledge-acquisition and understanding identifies a network involving association cortex, dorsolateral prefrontal cortex, orbitofrontal cortex, striatum, opioids, and dopamine. Candidate neural mechanisms are discussed in the Literature.

Let us repeat, the theory in Chapter 2, Mechanisms of the Mind: From Instincts to Beauty, predicts that aesthetic emotions correspond to changes in KI similarity. In addition to previously referenced publications supporting this prediction, further evidence is provided from pharmacological studies on the role of endogenous opioids in learning, retention, and memory. Studies in animal pharmacology suggest that the opioid receptor antagonist naloxone ($C_{19}H_{21}NO_4$; $M \approx 327.37$ Da) tends to *impair* retention and learning performances in rats. An exploratory study conducted by Goldstein demonstrated that blocking opioid receptors with the same synthetic molecule significantly decreased or inhibited chill responses. Schoeller and Perlovsky (2016) investigated the psychology of aesthetic chills in a series of experiments including its relation to knowledge-acquisition.

To summarize, based on the mathematical theory of the KI and aesthetic emotions, the following hypotheses were formulated and tested: (1) aesthetic chills correspond to an event where KI similarity between bottom-up sensory signals and top-down mental models' representations reaches a local maximum, whereas horror chills correspond to an event where KI similarity reaches a local minimum; (2) when KI similarity reaches a maximum near the top of the mental hierarchy, humans experience aesthetic emotions of the beautiful; some of these experiences are elucidated in this chapter, clarifying and defining the emotions of the beautiful; (3) coherent-meaningful prime enhances chill experiences, whereas incoherent prime inhibits chills; this indicates relations between chills and meaningful-purposeful contents; near the top of the mental hierarchy it relates emotions of the beautiful to meaningful-purposeful contents; and (4) aesthetic chills are significantly correlated with KI curiosity, participants who experienced chills had higher scores on the curiosity scale than subjects who did not, and curiosity significantly predicted whether subjects experience chills.

Additional details of the Method *(copied with shortcuts from the above publication). To test the model prediction that aesthetic chills are related to the meaning and purpose, we used a standard method in psychological tests, priming. Priming preactivates the corresponding system. It can be understood as a bottom-up neural activation by prior inputs which leaves a residual activity trace and makes it easier for the primed neural cells to be activated by similar future inputs. Our experiment uses primes that vary greatly in relation to the aim of the study - the meaning near the top of the cognitive hierarchy. The first prime (the meaningful prime) requires thoughtful cognition, the second prime (the neutral prime) requires no cognitive effort, and the third prime (the incoherent prime) has no meaning. Specifically, the primes are (i) a sentence from the French philosopher Pascal ('the supreme function of reason is to show man that some things are beyond reason'), (ii) a sentence from a children book ('I once saw an apple in a tree') and (iii) a classic example from Noam Chomsky illustrating that syntax might be independent from cognitive meaning ('colorless green ideas sleep furiously'). The Pascal prime encompasses the widest range of reality and might address representations near the top of the mental hierarchy.*

In a series of quantitative and qualitative studies we tested the relation between aesthetic emotion and meaning-making (hypothesis 2) using the methodology of priming, and tested the relation between KI and aesthetic emotions using a psychometric instrument (hypothesis 1). The phenomenology of the aesthetic experience was studied by asking subjects what it felt like to experience chills and, in order to gather material for further experimentation, and more importantly for relating chills to beauty and meanings, the properties of chill-eliciting narratives were studied by asking subjects after the experiment for a description of a scene in a narrative (film, book, play, etc.) that gave them the chills or made them shivers.

Quantitative study (priming and KI, curiosity). *To test the hypothesis the experiment used the discussed methodology of semantic priming. Subjects were divided into three groups, each primed with a different text. The treatment group (GROUP 1) is exposed to a coherent existential prime supposed to arouse their knowledge instinct by activating the categories at the top of the mental hierarchy (the models which encompass the widest range of reality and that are consequently highly difficult to manipulate by the subject). To remind, we used a quote from the philosopher Pascal: "the supreme function of reason is to show man that some things are beyond reason". Postexperimental interviews with subjects positively revealed that the prime was successful in engaging epistemic behavior at the top of the mental hierarchy and subjects in this priming group successfully experienced chills. GROUP 2 subjects were exposed to a neutral prime ("I once saw an apple in a tree") and we exposed the third group of subjects (GROUP 3) to an incoherent prime ("colorless green ideas sleep furiously"). Postexperimental interviews with subjects positively revealed that the prime was indeed rated by participants as "incomprehensible", "incoherent", "paradoxical", etc. Finally the amount of pleasure brought by the experience to the subject was quantified. This was done by using an analog rating scale (0 to 10). For chill-eliciting stimulation a film was used involving a dance sequence choreographed by Angelin Preljocaj and the adagio from Mozart's 23rd piano concerto, K488 (easily available online using the keywords, "Air France", "l'envol"). This stimulation is 60 seconds long and successfully elicited aesthetic chills. For further discussion on audiovisual contents more likely than others to elicit chills refer to Schoeller, 2015b. For evaluation and presentation of data, we used SPSS and Matlab.*

As a measure of KI-curiosity, we used Kashdan & Fincham's, 2004 measure of curiosity conceptualized as a positive emotional-motivational system associated with the recognition, pursuit, and self-regulation of novelty. It encompasses the two exploratory tendencies systemized by Berlyne: (i) diversive curiosity and (ii) specific curiosity. It also incorporates absorption, the tendency to become fully engaged in rewarding experiences. In other words absorption is the ability to ignore superfluous information and become absorbed in specific novel activities. For these reasons Kashdan & Fincham's measure of curiosity is appropriate for KI, and we refer to it as KI-curiosity.

Participants: *The experiment was conducted in Denmark, at a laboratory of the University of Copenhagen. A total of 30 international students at Copenhagen University from the Media, Cognition and Communication department participated in the experiment (16 females and 14 males; age range: 20–33). On their arrival at the laboratory waiting room the participants were randomly assigned to one of the three priming conditions.*

Procedure *(a short description): the participant entered the laboratory, sat in front of a blank screen in a comfortable chair. She was provided with a factual account of what is going to happen next. The participant was then asked to concentrate and is exposed to a prime that she had been randomly assigned to at arrival in the laboratory. The exact wording of the prime was as follows: "Please concentrate. Now, what do you think about this sentence? [priming]". After 90 seconds of prime, the subject was exposed to the chill-eliciting stimulus (a film excerpt). The stimulation was 60 seconds long. Once the experiment was finished, the subject received a questionnaire containing (i) Kashdan & Fincham's curiosity measure, (ii) demographics, (iii) an analog rating scale for the amount of pleasure (0 to 10) felt by the subject during the film (iv) narratological questions regarding chill-eliciting scenes in narratives and (v) various open phenomenological questions (such as "please describe in your own words what you have felt during the film that you have just seen", etc.). Each session lasted about 10 minutes.*

Qualitative study (narratology and phenomenology). *A qualitative data analysis was used to study two aspects of the problem: the phenomenology of the experience and the narratology of chill-eliciting scenes. To do so, we asked sixty students from the same population (thirty females and thirty males; age range: 20–33; Mean: 22.77) to describe a scene in a narrative (a film, a book, a play) that gave them aesthetic chills. The request was introduced with an open question: "Please describe as precisely as possible a scene in a film, a story or a play that gave you goose bumps (or chills, shivers, feeling of cold in your back, etc.)". We first analyzed the verbal reports paying attention to any redundancy that might be found in them. In what follows, we present these redundancies and the classes of situations eliciting chills using a very general narratological model, Polti, 1921. This model has the advantage of describing a wealth of scenes from different works of literature and films, which may be used for further experimentations on aesthetic chills in the laboratory, and describing the types of contents eliciting the emotions of the beautiful.*

RESULTS (SHORTENED)

The Beautiful, the Meaningful, and How They Relate

Quantitative Studies

Seven participants reported to have experienced chills in the course of the stimulation. Chills were nonrandomly distributed within the priming groups, none of them were found in Group 3 (the incoherent prime); this is a preliminary confirmation of the hypothesis: *aesthetic chills are related to the meaning.*

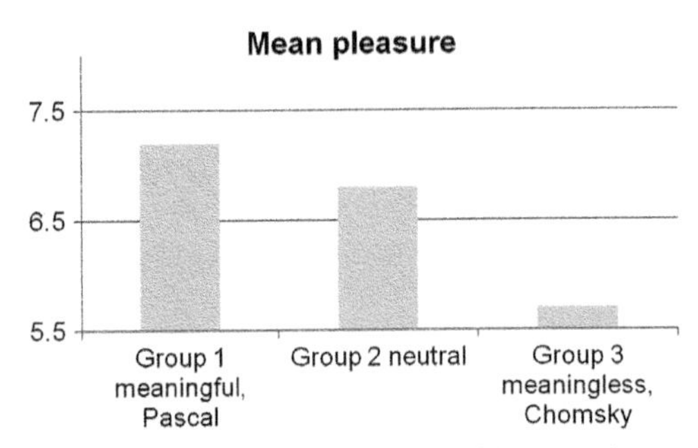

Figure 6.2 Participants primed with the meaningful (Pascal) sentence reported to have felt more pleasure than those primed with the meaningless (Chomsky) sentence. This supports the relation between aesthetic chills and emotions of the beautiful.

The difference in chill response between participants in Group 1 ($N_{chill} = 4$, $N_{group} = 10$) and participants in Group 3 ($N_{chill} = 0$, $N_{group} = 10$) is statistically significant ($p < .05$). *This confirms the hypothesis, the emotions of the beautiful are related to the meaningful-purposeful content of the experience.* (The accurate statistical formulation is that at the standard significance level this test *rejects* the null hypothesis: "occurrence of chills does not depend on priming.")

The analysis of the reports from the analog rating scale found a *significant difference* ($p = .03$) in *pleasure* between Group 1 and Group 3 (Fig. 6.2). This confirms that *the pleasure of the beautiful is enhanced by the meaningful-purposeful content of the experience.*

Curiosity significantly correlates in few ways with aesthetic chills. Correlation revealed that aesthetic chills were significantly, positively correlated with KI curiosity ($p < .01$). Participants who experienced chills had higher scores on the curiosity scale than participants who did not ($p < .001$). Curiosity significantly predicted whether subjects experienced chills ($p < .05$).

Qualitative Studies

Narratology

Our ultimate goal, let us remind, is to understand what is beautiful, what is meaningful, and to verify the theoretical prediction that these two are related. For this reason here we discuss the narratology of aesthetic chills: what structures and contents of stories elicit aesthetic chills. Table 6.1 summarizes chill-eliciting films and genres.

Table 6.1 Chill-eliciting films and genres

Film	Country of production	Year of production	Genre
Saving Private Ryan	United States	1998	Drama
Titanic	United States	1997	Drama
Avatar	United States	2009	Adventure
Todo Sobre su Madre	Spain	1999	Drama
Matrix	United States	2003	Science-fiction
Twelve Years a Slave	United States	2013	Drama
Panic Room	United States	2002	Drama
Good Will Hunting	United States	1997	Drama
Derek	United Kingdom	2012	Drama
Helium	Denmark	2014	Drama
Hunger Games	United States	2012	Adventure
Harry Potter	United States	2011	Adventure
Lion King	United States	1994	Adventure
The Notebook	United States	2004	Drama
Across the Universe	United States	2007	Drama
Whiplash	United States	2014	Drama
Dark Knight Rises	United States	2012	Thriller
The Notebook	United States	2004	Drama
Taken	United States	2008	Thriller
The Bridges of Madison County	United States	1995	Drama
Dead Poet Society	United States	1989	Drama
Interstellar	United States	2014	Adventure
The Others	United States	2001	Thriller
Green Mile	United States	1999	Drama
The Beach	United States	2001	Adventure

Properties of chill-eliciting scenes: More than 80% of the chill-eliciting scenes described by the experiment participants take place in the third act of an American film and involve music (see Table 6.1). Two film genres are largely predominant: drama films and adventure films (IMDB's classification). Chill-eliciting scenes usually involve radical changes in the *relations* among characters; e.g., separation through death or reunion after strong and demanding efforts (here 62% of our scenes). These relations are usually of three kinds: (1) social relations, (2) family relations, most notably parents-to-children relations, and (3) couple relations. In case of a reunion scene, the amount of pleasure seems to be proportional to the quantity of effort necessary for the reunion. Chills can also be enhanced by a strong change in the affective display of one of the characters (when empathy is a necessary condition to reduce narrative tension to a minimum). Chills are also experienced when characters are no longer acting for their own sake but for a cause greater than themselves. As when a character is lying to others in order to protect them (i.e., hiding something fundamental for himself but harming for others) or when people sacrifice themselves to something extraneous. That is, when the values driving a given action are stronger than the likelihood for this action being successful (which are usually null, as in sacrifice, heroism, bravery toward the impossible, etc.). Another important factor is the disclosure of the plot (58% of our sample), chill-eliciting scenes are usually a crucial node in the narrative which they belong to—i.e., the summary of the film would not be the same if the properties of these scenes were to be different. In other words, *chill-eliciting scenes reveal the meaning of the story.*

Classes of situations eliciting chills: The 36 Dramatic Situations is a descriptive list (a narrative model) created by Georges Polti to categorize every dramatic situation that might occur in a narrative. His work was influenced by Goethe and Carlo Gozzi. For each situations, Polti details the "indispensable elements" and provides extended explanations and examples. Examples of situations could be entire work of literature (e.g., Aeschylus' *Suppliants*), parts of a work (the third act of Voltaire's *Tancred*), historical examples (a historical episode of the crusades) or situations of ordinary life (e.g., claiming the body of a dead relative in the hospital). Here, we present the various situations which were found in our sample, provide the examples offered by Polti and add films quoted by participants.

1. DELIVERANCE (Polti's 2nd situation): involves an unfortunate, a threatener, a rescuer; in which an unexpected protector, of his own accord, comes suddenly to the rescue of the distressed and despairing. The first two subclasses of this situation are found in our sample of scenes:
   a. Appearance of a rescuer to the condemned: Sophocles' "Andromeda"; the first act of "Lohengrin," the third act of Voltaire's "Tancred"; the denouement of "Bluebeard." In our sample: Panic Room (Fincher, 2002).
   b. A parent replaced upon throne by his children: most notably "Aegeus" and "Peleus" by Sophocles; Euripides's "Antiope." In our sample: The Lion King (Walt Disney Pictures, 1994).

2. DISASTER (Polti 6th situation): involves a vanquished power, a victorious enemy or a messenger. This situation is best defined by a great reversal of roles, the powerful are overthrown, the weak exalted. Most notably
   a. A.2 A fatherland destroyed as in Byron's "Sardanapalus" and "The War of the Worlds" by Wells. In our sample: Hunger Games (Ross, 2012).
   b. C.1 (Ingratitude suffered), sacrifice unappreciated by those who have benefited by it, as in the "Timon of Athens" and "King Lear" of Shakespeare's. In our sample: Whiplash (Chazelle, 2014).
3. ADVENTURES (Polti's 9th situation): involves a bold leader, an object and an adversary. Two subclasses of this situation are noticeable in the descriptions provided by our subjects:
   a. Preparations for war: in this subclass, the action stops before the denouement which it leaves to be imagined in the perspective of enthusiastic prediction, e.g., Aeschylus' "Nemea;" "The Council of the Argives" by Sophocles. In our sample: The Dark Knight Rises (Nolan, 2012), Avatar (Cameron, 2009).
   b. Adventurous expedition: the usual exploits of the heroes of fairy tales. Euripides' "Theseus"; Sophocles' "Sinon." The Labors of Hercules; the majority of Jules Verne's stories. In our sample: Interstellar (Nolan, 2014), The Beach (Boyle, 2000).
4. SACRIFICE (Polti's 20th and 21st situations). There are mainly two types of sacrifices quoted by our participants:
   a. Self-sacrificing for an ideal which involves a hero, an ideal and the "creditor" or the person or thing sacrificed. Here, Polti notes that 'the field of conflict is no longer the visible world' (1921, p. 67). It finds familiar instances in all martyrs. In fiction: 'L'Œuvre' by Zola and 'L'enfant du Temple' (de Pohles). In our sample: Saving Private Ryan (Spielberg, 1998).
   b. The second type of sacrifice is self-sacrifice for kindred which involves the hero, the Kinsman and the "creditor or the person or thing sacrificed." Most notably C(2) Sacrifice for the happiness of ones child—e.g., "Le Réveil" (Hervieu, 1905); "La Fugitive" (Picard, 1911) ; in our sample: Titanic (Cameron, 1997).

Phenomenology of Chill Episodes

This subsection is summarized in Table 6.2.

Table 6.2 Phenomenology of chill episodes

Effect	Description	Quotes
Relaxation Psychologically, similarity (L) reaches a maximum and its rate of change approaches zero (Schoeller & Perlovsky, 2015). Physiologically, midbrain dopaminergic neurons reward the system (Blood & Zatorre, 2001; Goldstein, 1980).	Most subjects mention a strong relaxation and a calming effect. We note that some mention both curiosity and relaxation.	• Something calm and soothing that made me relax. • Was intrigued and calm at the same time. • Soothed and relaxed. • I felt far more relaxed, almost enchanted. • I was quiet, relaxed. • Its very calming. • Peaceful and curious at once. • Relax and curious of what was going to happen. • I felt very relaxed. • I felt a sense of peace and quiet.
Neutrality Psychologically, derivatives of L tend toward zero, thus emotion is experienced as neutral.	In terms of emotional valence, it seems as chill episodes are neither positive nor negative.	• Neither a positive nor a negative moment, but a moment through which you don't belong to yourself anymore. • You can't divide the feeling on a dualistic level and say that it's either painful or pleasant.
Psychological impact Categories at the top of the mental hierarchy were positively reinforced, highest models have been modified, the entire architecture of the cognitive system proved functional, something important was learned.	Subjects report a strong experience.	• It indicates that the movie has some sort of impact on me. • It was a very emotionally strong scene, and I like to get touched in such ways. • That always evokes a strong feeling of freedom in me. • The moment they started to dance seemed very emotional and dramatic. • It forces you to reconsider things.

Arariskein[a]	Subjects describe some sort of concinnity effect, the harmonious adaptation of parts into a coherent-meaningful whole.	• It made me focus on the bigger picture rather than the details of the images because of the atmosphere I was presented with. • The part where the small streams run together to form the river will send chills down my spine anytime [describing Smetana's - ma vlast Vltava]. • It was like watching a film where one person forces another to aggregate in order to produce a greater and more powerful whole.
Empathy	Subjects report an unusual strong empathy for the characters.	• I could say identification with the character. • I could feel the passion of the couple presented. • I felt carried away by the dancing couple. • When you connect with the character as a result in a film following them through their exploits, and a moment impacts upon you as much as them. • I kind of reacted the same way as the girl in the film. • Empathy for an emotional moment of the singer. • Could almost feel their sorrow myself.

(*Continued*)

Table 6.2 (Continued)

Effect	Description	Quotes
Authenticity	Subjects talk about authenticity	• The end was disappointing, it looked like an ad and this destroyed the authenticity. • Disney movies and cartoons can give me goose bumps. But not romantic comedies. It has to seem authentic. • These films I guess ultimately make you question what is real in real life.
Aesthetic criterion	Subjects intentionally seek for this kind of moments in films, they sometimes use it as an aesthetic criterion.	• I would always seek it again. • I think it makes a good movie, song, etc. • From having chills, I know that I like the film
Miscellaneous The manipulation of the category at the top of the mental hierarchy is mostly unconscious (since difficult to manipulate and risky in terms of error—modification of the entire world of the subject would follow) but the emotion seems conscious.	Note that people do not seem to be reflecting so much about chills on a day-to-day basis, they usually do not know how to account for them and do not pay much attention to the event as such. They are surprised (and curious) when we bring the phenomenon to their attention.	In postexperimental interviews many subjects stated that they "never wondered about it, never thought of that before," etc.

Quotes are copied from students without editing.

[a]This untranslatable term derives from the ancient Greek ἀραρίσκω (ararískō, "to fit together") and can be more or less equated to the Gestalt effect, the cognitive system's ability to create forms to simplify raw perceptual data.

DISCUSSION

Quantitative Data

Experimentally, the working hypothesis is tentatively confirmed by the present results which verify that (1) aesthetic chills significantly correlate with KI curiosity, (2) exposing subjects to an incoherent prime strongly decreases the likelihood of the subject experiencing chills during the subsequent exposition to a chill-eliciting stimulation, and (3) exposing subjects to a meaningful prime prior to the chill-eliciting stimulation makes the participant's experience more pleasurable. These demonstrate the relation between the beautiful and meaningful.

The experimental protocol, is replicable, therefore allows for a scientific study of both aesthetic emotions and meaning.

Qualitative Data

Narratology

To understand the structure of the scenes described by participants in our experiments and functions of these scenes in the narratives that they were extracted from, we would introduce now general principles of narratology. Since Aristotle, it is generally accepted that narrative structures are a representation of human action (etymologically, drama means action). Human action is usually described as organized hierarchically. Each action is a step toward fulfillment of a plan, which is itself a contribution to a more comprehensive plan, and so on, until overriding plans, corresponding to life goals, are reached. Thus, ultimately, a successful story, just as a popular myth, should provide its audience with a better understanding of human action at the most general level; this corresponds to the hierarchical KI theory in Chapter 2, Mechanisms of the Mind: From Instincts to Beauty.

Most of the scenes described by our subjects are extracted from Hollywood drama narratives. These follow a specific pattern largely influenced by Aristotle's Poetics and one simple model describing this type of narrative structure is found in various manuals for storytellers widely used by contemporary scriptwriters. The most basic elements of the structure are hero, goal, obstacle, and conflict. These elements are tied together by what we may refer to after Kreitler and Kreitler (1972) as the tension principle. The hero has a goal, he encounters obstacle, this creates tension (conflict) and the film ends when an answer to the dramatic question has been given (will the hero reach his goal?).

Hollywood narratives can thus be described in terms of retention of information. The answer to the dramatic question is the information that the author retains. It is what the viewer cares about, what motivates his curiosity. As a consequence, if the audience suspects that the author does not hold such information, the spectatorial contract (i.e., "I am going to be told a story which end I shall know") is broken, the tension principle ceases to apply, and the viewer has no good reason to wait until the end of the story. This, among many other things, is one of the reasons why one does not interrupt a theater play; he wants to know the end *as it will happen tonight* and if he interrupts the play, he will *never* know such end.

This highly simplified explanation allows for a mathematical description outlining possible extension of our Chapter 2, Mechanisms of the Mind: From Instincts to Beauty, theory to dynamical representations. The hero (e.g., Bobby) has a goal G (e.g., he wants to become the best chess player in the world) but there exists an obstacle O (e.g., he does not play sufficiently well to beat his rival Garry). The amount of conflict, c, is inversely proportional to the likelihood of Bobby reaching his goal G given obstacle O (e.g., the less knowledge of chess, the greater the difficulty for Bobby to beat Garry at chess, the greater the amount of conflict). A story is the depiction of the process of Bobby acquiring enough knowledge to become the best chess player in the world. We may thus write the following structural equation:

$$c = 1/P(G \mid O)$$

Here, c is a conflict and $P(G|O)$ is the conditional probability of Bobby reaching his goal G given obstacle O. The lower the likelihood of success, the greater the narrative tension. This indeed is a highly simplified model. There are many characters in a given story and their goals are usually conflicting. Models are metaphors, they serve to explore and explain processes not yet accessible to observation. We can use them to build testable predictions. From this highly simplified model, it seems possible to distinguish between at least two different types of stories. Stories where the conflict is function of the hero wanting to achieve a general goal, e.g., money, fame, power, and stories where the conflict is function of the hero wanting to avoid a general obstacle, e.g., sickness, catastrophe, misery. Stories involving conflicts of the first kind (internal conflict) have a different structure and trigger different emotions than stories involving conflicts of the second kind (external conflict). Narrative tension is directly proportional to conflict and is inversely proportional to the likelihood of the hero reaching his goal given obstacles.

All of the scenes described by our subjects involve the main character of the story that they were extracted from. The main character of a given story is generally defined as the hero that suffers the greatest amount of conflict. There exist mainly two types of heroes: (1) heroes who have tremendous goals, e.g., being the best chest player in the world and (2) heroes who have tremendous obstacles, e.g., surviving a natural catastrophe. Aristotle noted that there are mainly two types of tragedy both involving a great reversal in fortune. The first kind is a reversal from good to bad fortune (e.g., *Oedipus*) the second kind involves a reversal from bad to good fortune (e.g., *Eumenides*). According to him, the second is more pleasurable as it involves pity and fear and both are fundamental for *katharsis*, the *function* of tragedy. Most of the scenes described by our subjects fall under the second category (Good Will Hunting, Titanic, The Notebook, All About My Mother, Twelve Years a Slave, Panic Room, etc.).

If the story is well-crafted (if the classical unities are respected, if the elements of the narratives are important, if the viewer cares about the story) narrative tension should trigger psychological tension, i.e., an emotional response. In the course of the story, tension increases and decreases. The reduction of tension causes pleasure, positive emotions, whereas the increase of tension causes displeasure, negative emotions. There are particular nods in the story, especially at the end, in the third act, where tension reaches a peak value. These nods constitute the structure of the story. They are usually contained in the simplest and shortest description of the story—i.e., its meaning. Popular blockbusters are often very easily summarized in a few sentences and such is the case for most films in our sample. Aesthetic chills seem to occur after tension reaches a peak value. They occur in the last third of the plot when narrative tension reaches a global minimum. Paradoxically, if we consider the properties of the scenes they also seem to involve an important conflict (thus the summary of the film would be different if the properties of these scenes were to be changed—it would not be the same story). The dynamic of our scenes seems to involve a change from a great disequilibrium ($P(G|O) \sim 0$) to a state of equilibrium ($c \sim 0$). Consider, e.g., the following scenes described by our participants (we do not change participants wordings):

- Panic room: there were three burglars entering the house of a rich family: the family was in an safe room which was supposed to be for cases like this: the daughter was suffering from asthma and almost died at the end: at the end one of the burglars shot the two other ones in order to save the daughter.

- The Bridges of Madison County: when Meryl Streep's character looks back at the man she loves but can never be with. It made me feel sad and disappointed with the world.
- The Imitation Game: when they solve the Enigma code and find out that one's brother is on board of one of the ships being bombed by the Germans, but there's nothing they can do about it, because the enemy will get suspicious. They just all stand, looking at the map with tears in their eyes.
- The final scene of the film "12 years a slave" at the point when the main character (who is in silence for the majority of the film) is reunited with his family after his 12 year ordeal. He stands opposite to his wife and two children and their families. In this moment we see the life that he has missed. Very few words are said, initially it is a cagey and uncertain affair, the audience feels maybe too much has happened for any of the characters to be capable of understanding one another, but the tension ends with an outpouring of emotion and embracing. This is made more intense for the audience because we have seen how the main character has endured so much, but in this moment the weight of everything is released, he is softened and can once again feel.
- The scene in *Titanic* when the ship is about to sink and the camera focuses on a mother lying in bed with her children. At that point it is already clear that they will not being able to leave the ship before it sinks. Still, the mother is telling them a goodnight story.

The most remarkable thing to notice here is that chill-eliciting situations are not situations of conflict per se. These scenes do not involve tension since the probable occurrence of a relevant event is not an undefined course of events. They do not provide clues for the viewer to predict future events. That is, the viewer's inferences find their conclusions in the properties constituting the scenes themselves (for further discussion refer to Schoeller & Perlovsky, 2015). There is no need to foresee what is going to happen next since the storyline has reached its point of equilibrium. The spectator is finally free to let go, the forces in the conflict field are balanced. At the narratological level, equilibria play an essential role. Chill-inducing scenes take place in the last third of the films; what Aristotle refers to as the third act. Freytag talks about denouement, the Russian formalist Vladimir Propp talks about "the hero's return" and

Todorov "a final equilibrium." By the end of most of the scenes from our sample an answer to the dramatic question has been provided. This indeed corroborates the idea that disclosure of the plot seems to be a strong elicitor for aesthetic chills. (Note: here the subjects describe strong aesthetic emotions related to the meaning of the story, in other words near the top of the mental hierarchy; these are the emotions of the beautiful! The beautiful is defined not by particular shapes, or colors, or gestures, or rhymes, it is defined by highly meaningful-purposeful content). According to Hegel, tragedy emerges when a hero asserts a justified position, but in doing so simultaneously violates a contrary and similarly justified position.

> *The original essence of tragedy consists then in the fact that within such a conflict each of the opposed sides, if taken by itself, has justification, while on the other hand each can establish the true and positive content of its own aim and character only by negating and damaging the equally justified power of the other.*
>
> ***Hegel (1975)***

As noted by Roche, the nature of the tragic hero is therefore paradoxical as "greatness comes at the price of excluding what the situation demands" (the burglar should not save the daughter, Meryl Streep should not let her love go, the mother should not tell a story to her kids, the character in the *Imitation Game* should not sacrifice his brother—but, they are *forced* to do otherwise). In the words of Hegel, "it is the honor of these great characters to be culpable" and the tragic heroes embrace conflicting positions that are "equally justified." Here, it is possible to draw a parallel between Hollywood narratives and mythic structures in general. The structuralist theory of myth considers that the purpose of mythic structures is to provide a logical mode capable of overcoming contradictions. The greater the contradiction resolved, the more powerful the myth.

We would like to mention that the classes of situation eliciting chills are closely related to the usual elements of propagandistic materials (sacrifice, martyrdom, preparation for war). More empirical evidence is provided by the only other study investigating narrative chills (Konecni et al., 2007). Given the current danger posed by such material, we encourage researchers in the field of propaganda and security studies to investigate chill-eliciting scenes and their psychological counterparts. Succinct though essential details regarding our database were provided, we encourage researchers to build their own and pursue the analysis further.

Studies in social cognition have shown that cognitive processes are heavily dependent on antinomies and polarization and so one might postulate that the universal success of Hollywood narratives and their general propensity to elicit chills might be due to their capability to help overcoming some of the most robust cognitive antinomies by presenting playful solutions to fundamental human conflicts. Following Festinger, we would thus propose that chill-eliciting scenes provide viewers (readers) with new cognitive elements that maintain the *total dissonance of the system* at a rather low level, thus allowing the subject to divorce himself psychologically from the conflict.

The basic background of Festinger's theory of cognitive consonance is coherent with the KI maximization model of cognition proposed in Chapter 2, Mechanisms of the Mind: From Instincts to Beauty, Chapter 3, Language and Wholeness of Psyche, and Chapter 4, Music, and consists in the notion that the human organism strives to establish internal harmony, consistency, or congruity among his opinions, attitudes, knowledge, and values. However, depending on cultural or biological determinants, there exist circumstances that naturally lead to fundamentally disharmonious, inconsistent, and incongruent cognitions. These opposite pairs of cognitive elements (by-products of knowledge-acquisition) remain infinitely resistant to change, this robustness being due to their irrevocability. The possibilities of change for such cognitions are almost nil since there is a clear and unequivocal reality corresponding to them. Empathy seems to play a crucial role in these scenes (e.g., in many of the descriptions, empathy is a necessary condition to reduce narrative tension to a minimum). And of course the role of a creative author is to identify such cognitions, which have not been noticed previously, and to find previously unknown resolutions. To identify the kind of biological or cultural conflicts responsible for these opposite pairs of cognition, we must turn our attention to the phenomenology of the experience.

Phenomenology

Our results indicate that viewers experience a strong relaxation during and after the chill episode and that chill-eliciting scenes have some sort of calming effect. Aesthetic emotions function in the management of goals, KI maximization. It is positive when the goal is advanced and KI increases, negative when the goal is impeded. As we already discussed chill-eliciting scenes are reported in two film genres in particular: drama films and adventure films. Films from these categories involve complex situations of tensed conflict and thus are more prone to elicit chills since they are the center of very wide conflict fields. Coherent with the narratological analysis, it is possible to draw a parallel here between these narratives and mythic structures in general. The structuralist theory of myth let's repeat, considers that the purpose of mythic structures is to provide a logical mode capable of overcoming contradictions. We saw that the universal success of Hollywood narratives and their general propensity to elicit chills might be due to their capability to help overcoming some of the most robust cognitive antinomies by presenting *playful* solutions to fundamental human conflicts. Chill-eliciting scenes provide viewers (readers) with new cognitive elements that maintain the total dissonance of the system at a rather low level, thus allowing the viewer to divorce himself psychologically from the conflict.

Continuing the discussion of Festinger's and KI theories, from the end of the previous subsection one page up, the strong and unusual empathy mentioned by our subjects leads us to propose that one of the fundamental human conflicts which chill-eliciting scenes might help to resolve and overcome is related to the primordial need in others, humans survive by sharing goals but can never access other's thoughts directly (nobody will ever have direct access to your thoughts and conversely you will never have direct access to somebody else's thoughts). A related fundamental conflict which chills might help overcome is the altruistic nature of human beings as opposed to the high degree of egoism present in contemporary cultures. The narratological results seem to suggest that scenes eliciting chills do so by displaying situations where empathy is necessary to reduce the narrative tension to the minimum. I would suggest that the best narratives in history have extended understanding of empathy current to their times toward ever increasing possibilities for human individual consciousness. More empirical research would prove useful in improving our understanding of these processes.

CONCLUSION

This chapter is a step toward experimental confirmation of theoretical predictions made in Chapter 2, Mechanisms of the Mind: From Instincts to Beauty, about emotions of the beautiful, representations of the meaningful and purposeful, and their relations. The contents of the beautiful, meaningful, and purposeful has been elucidated, as well as their relationships. One has to remember that contents of these highest representations is not directly accessible to consciousness.

The theory that has been experimentally tested in this chapter is in line with a philosophical tradition that has long associated aesthetic emotions to knowledge, this theory proposes that aesthetic emotions correspond to a satisfaction/dissatisfaction of humans' KI, the primary drive to acquire knowledge about the external and internal world and perceive events as meaningful. It is coherent with the known biology of both aesthetic chills and knowledge-acquisition. I remind, aesthetic chills were found to be significantly decreased or inhibited by injection of the synthetic opioid-antagonist naloxone and laboratory studies in animal psychology provide extensive evidence that learning is influenced by opioid peptides. Furthermore, both knowledge-acquisition and aesthetic chills activate striatal regions involved in reward processing and coding for vital parameters. This chapter summarized the first investigation into the relation between aesthetic chills and knowledge-acquisition.

According to the theory described in Chapter 2, Mechanisms of the Mind: From Instincts to Beauty, aesthetic experience involving the content at the top of the cognitive hierarchy (*the highest meaning*) corresponds to an event when the mental representations encompassing *the widest range of reality* are positively reinforced by external cues, when the entire architecture of the cognitive system encounters supportive evidence that it is grounded on perennial and appropriate foundations. This rewards the individual with pleasure (*the emotions of the beautiful*).

Such events, because they correspond to a change in similarity knowledge, may vary in *degree* proportionally to the importance of the cognitive elements at play. The strength of emotion depends on the role played by these elements in the architectonics of the total system and the strength of the occurring response is a function of the amount of mismatch between bottom-up and top-down configurations. The highest levels of the hierarchy correspond to the most general and abstract mental models and their modification induces changes at all lower levels. Hence, such modifications are highly demanding in terms of resources, risky in terms of error and the content at the top of the hierarchy is highly difficult to manipulate consciously by the individual which might explain why people do not reflect about chills on a day-to-day basis, why they are surprised when we bring this phenomenon to their attention. It resists change since an alteration of the subject's entire world would follow. These "highest" aesthetic experiences involve representations near the top of the mental hierarchy, which cognitive contents we experience mostly unconsciously as "meaning of life." Acquiring knowledge about these representations even unconsciously invoke the emotion of the beautiful. Here we are coming to understanding the nature of the beautiful not through a specific content, such as the visual features of a painting, but rather as what makes one's life meaningful.

This theory has demonstrated major explanatory power as discussed throughout the book. This chapter established a relation between aesthetic chills and the attribution of meaning. First, by showing that chills occur in a group exposed to meaningful prime, while an incoherent prime acts as a strong inhibitor for aesthetic chills, and second, by establishing that chill-inducing moments relate to meaning-making contents, content that is psychologically relevant for the participant. Though further experimentation should examine these relations in more detail, in current experiments our hypothesis has been tentatively confirmed, and we conclude from these results that aesthetic chills might correspond to a satisfaction of our experiment participants' vital need for knowledge and related to the beautiful and meaningful. A neural theory of how pleasure related to knowledge-acquisition and understanding arises from a network involving association cortex, dorsolateral prefrontal cortex, orbitofrontal cortex, striatum, opioids, and dopamine is outlined in Levine (2012). These detailed neural mechanisms of generalized mental representations are not yet accounted for in a simplified mathematical model of interacting emotions and cognition serving as a foundation for Chapter 2, Mechanisms of the Mind: From Instincts to Beauty. An important aspect of these findings is that our experimental protocol allows for a scientific study of the physiology of aesthetic chills. Given that exposition to an incoherent prime strongly inhibits chills (as opposed to, e.g., the sentence from Pascal), it is now possible for the experimenter to have *two groups of subjects*, both exposed to *the same stimulation* where subjects in one group experience chills while none of the other do so. Based on these results, we would like to suggest a series of propositions and hypotheses for further research on the problem of aesthetic emotions.

1. This study should be replicated in a different sociocultural setting and with physiological measurements. As underlined throughout the chapter, our experimental protocol can be helpful to determine the physiology of the phenomenon. Given that incoherence seems to be such a strong inhibitor, an electrophysiological study of the problem of chills and the role of signals often related to the activity meaning-making

(e.g., N400) in the activation/inhibition of chills might prove useful and shed some light on fundamental aspects of human nature (the relation between cognition/recognition, emotion/cognition, curiosity/pleasure/learning). Such a study if added to the appropriate measurement tools might also clarify the problem of coherence and meaning. A physiological study could prove useful to verify the empirical findings of Goldstein (1980) and Blood and Zatorre (2001). It could also be of interest to determine whether ceremonial chills, musical chills, narrative chills, artistic chills, scientific chills, and religious chills correspond to the same biological mechanisms and if not what are the fundamental differences.

2. Consistent with Berlyne's classic psychological theory of curiosity and available data concerning aesthetic chills, biological studies of curiosity seem to demonstrate that curiosity activates brain regions sensitive to conflict and activates striatal regions involved in reward processing. These studies also revealed a positive correlation between curiosity and memory retention; this might explain why participants mention a moment of psychological importance. Representations near the top of the hierarchy are mostly inaccessible to consciousness, the theory suggests that the category "meaning of life" is at the top of the mental hierarchy and unifies the entire cognitive system. Priming the subject with this specific category we used as a way to test this hypothesis. It should be done below the threshold of consciousness, as a participant will probably react antagonistically if exposed to the category consciously (given the very high degree of generality of the category and its unclear purpose, it is easier for a participant to dismiss the information rather than giving it its full attention). We hypothesize that the greater the attentional resources allocated by the subject on the primer, the stronger the emotion.
3. We also advanced the hypothesis that aesthetic chills might be causally related to terrorist indoctrination. Participants in our studies repeatedly mentioned that the notion of sacrifice seems to play a causal role in the elicitation of chills and the classes of situations eliciting chills often correspond to those of propagandistic material at large. This is coherent with previous results. Studies in general sociology show that when related to chills, the notion of sacrifice also plays a causal role in religious conversion. We strongly encourage researchers specialized in security studies to pursue the analysis further by investigating the relation between humans' vital need for (re)cognition, biological altruism, aesthetic chills, horror chills, and propaganda material.

4. There does not exist any study on the ontogeny of chills (at what age do subjects start experiencing them? is there a relation between frequency of chills and age?) Nor on the evolution of chills with cultures (how do chill-eliciting stimuli evolve over time? are there cultural differences? are there mother-language differences, as discussed in Chapter 3, Language and Wholeness of Psyche, and references therein?). Other studies have shown that chill-eliciting stimuli at large were found to cause an improvement of mood. We also proposed that chills may help overcome some fundamental cognitive conflicts, pairs of cognition of equal resistance to change where the least resistant element may not be altered nor modified. We proposed two such conflicts: the vital need for other versus cognitive solitude and biological altruism versus cultural selfishness.

   Following Festinger's original intuition, we proposed that chill-eliciting stimuli might provide subjects with *new cognitive elements* maintaining the total dissonance of the cognitive system at a low level. It was hypothesized that Hollywood films do so by displaying playful solutions to fundamental human conflicts. Berlyne (1949) noted that for knowledge of any given outcome to be rewarding, the event must be of some "interest" to the subject, it must be important, it must be meaningful—he equated this with the idea that strong habits or drives must be aroused. This is coherent with various accounts of the function of myths and narratives in human life: "unless the plot of the dream play represents a solution to a conflict in his spectators' life, the dream artist will have no audience at all, and hence cannot receive a social sanction for his play. On the other hand, if the plot of his play treats problems of great moment for his spectators, and treats them to their satisfaction in the sense that the spectators become clearer about the nature of their problems and of their solutions, then the theater may well become filled."

5. It is interesting to note that some subjects relate their experience to immersion in a natural environment (e.g., beach, forest, mountain). The restorative value of nature as a vehicle to improve cognitive functioning has been the object of various studies in recent years. Exposure to restorative environments facilitates recovery from mental stress and fatigue (Berto, 2005). The attention restoration theory postulates that interacting with environments rich with inherently fascinating stimuli (e.g., sunsets) invoke involuntary attention moderately, allowing directed-attention mechanisms a chance to replenish.

According to these studies, natural environments minimize the requirement for directed attention and therefore seem to have a restorative effects on cognitive functioning. After interacting with nature, participants in experiments are able to perform better on tasks that depend on directed-attention abilities.

Cognitive enhancement is also one of the known psychophysiological effects of music upon their listeners. In this book it is related to the fundamental function of music, overcoming cognitive dissonances. More research on the relation between the aesthetic experience and nature interactions would be required before postulating any therapeutic outcomes, but if such a relation were to be clearly established, this could open a promising future for the study of aesthetics and the therapeutic implications of aesthetic chills.

6. The role of coherence in art appreciation has been well studied. However, it is not yet completely clear why *incoherence* is such a strong inhibitor for aesthetic chills. In further research, it would be appropriate to study the effect of an incoherent prime on other universal emotions. Are subjects exposed to an incoherent prime prior to a stimulus likely to elicit sadness, e.g., a mother separated from her child, or fear, e.g., snakes, heights and other evolutionary relevant stimuli, less prone to react emotionally? if so, would the effect observed be as strong as the one observed in our experiments? Since, as underlined in many of our participants' reports, *harmony* seems to play a crucial role in the elicitation/nonelicitation of chills, it might also be useful to replicate this study on specific populations (e.g., musicians and mathematicians) and vary the prime in accordance (e.g., systemic integrity of a musical piece and elegant axiomatic of geometry). Such experimentations, if coupled with appropriate measurement tools of the physiology at play, could prove useful for determining if chills as elicited by audiovisual content pertains to the same class of biological phenomena as chills elicited by purely auditory or purely visual content. It could also shed some light on the psychobiological similarities and differences of chills as elicited by art, by science, and by religion and on their respective roles in modern societies.

CHAPTER 7

Music and Culture: Parallel Evolution

Contents

Abstract

Fascinating stories trace connections of consciousness and music from ancient world to the 21st century. This evolution is a struggle between differentiation and synthesis; it requires unified actions of language and music. From millennial history I select few episodes. In ancient Greece Apollonian and Dionysian consciousness is related to music from Homer to Attic tragedies. In the language of contemporary science I discuss Nietzsche's comparison of tragedy to musical dissonance, and the death of tragedy in Socrates' differentiation. Parallel evolution of consciousness and music is traced in ancient Israel from King David, Amos, and Seraphim antiphon in the vision of Isaiah to the last prophet Zechariah, at the same time when in Greece the first philosopher Thales pronounced "know thyself." Contemporary consciousness emerges in parallel with Nehemiah's antiphon. I trace synthesis and differentiation in consciousness and music during Early Christianity and Middle Ages; invention of musical notations and polyphony; love songs of troubadours and trouveres. I relate evolution of music during the Renaissance and Reformation including Baroque music to individuation of consciousness, music of Rococo and Classicism to rational consciousness, music of Romanticism to consciousness split by differentiation of the idea of rationality. Developments of consciousness in the 20th and 21st centuries away from the individual and toward the objective, collective socialism are related to dodecaphony, serialism, Modernism, postmodernism, minimalism, to elimination of emotions from music. Differentiation of the objective pushes against each other opposite tendencies in cognition and music. Parallel evolution of cognition and

Music, Passion, and Cognitive Function.
DOI: http://dx.doi.org/10.1016/B978-0-12-809461-7.00007-8

music is traced in tonal organization to the down of human evolution to before 250,000 BCE.

EMPIRICAL EVIDENCE IN HISTORY

Previous chapters discussed the theory of the fundamental role and function of musical emotions in cognition and evolution of culture. This theory is supported by laboratory experiments.

Now I review the empirical support for this theory. I consider historical evidence for the parallel evolution of culture, consciousness, and musical styles. The idea of consciousness changing through time might be unpalatable for some people. The difficulty is psychological: everything we know immediately about ourselves is simply the current contents of our consciousness. Everything else requires concentrated analysis. By looking at the people around us and remembering growing up from childhood to adolescence, we easily notice that the contents of our consciousness have changed significantly.

Some ideas in this book relating to evolution of music, culture, and consciousness came from Nietzsche (1876). Jaynes (1976) analyzed historical changes in the contents of consciousness by studying preserved texts, including the Bible and Homer. Independently, Weiss and Taruskin (1984) and Perlovsky studied changes in musical styles. This evidence demonstrates that advances in consciousness and cultures paralleled advances in differentiation of musical emotions. Here we select a few examples from this history. But before going to concrete examples, let us recollect the main theoretical arguments considered in the previous chapters.

Review of Theoretical Arguments

Let us recollect the main theoretical ideas. The interaction between differentiation and synthesis is one of the general laws of how knowledge instinct (KI) operates, characteristic of any epoch in human history. Differentiation counteracts synthesis, enlivened by music. Uniqueness of music among other arts is in its synthetic effect on psyche, directly involving unconscious and emotional. This produces creative state of a collective soul and moves the cultural process. Condition of synthesis is a correspondence of concepts of everyday material life and the highest spiritual purpose. In such a state, culture develops and the power of spiritual gaze strengthens, spaces surrounding man move apart and distances open, new more diverse concepts and emotions enter into the field of regard.

Accelerated differentiation of everyday life can tip the balance between the everyday and the meaning and purpose of life. "It is difficult to keep the scissor blades together." It is difficult indeed because the condition of the creative process involves the combination of opposites, differentiation and synthesis. Their complex dynamics determines the development of culture. When unity within the soul is achieved (synthesis), creative energy is directed at exploration of the outer and inner worlds, at widening the sphere of the conscious—i.e., diversification-differentiation of everyday concepts and emotions. So, Judeo-Christian synthesis prepared the ground for the understanding that man is the source of creative spirit, and this groundwork formed the conditions necessary for the emergence of scientific thinking, although it took thousands of years to come to fruition. It was not until the 17th century that Descartes completed his "expelling spirit from matter" and Newton, following him, could think about the completely causal, that is scientific, explanation of the material world.

In the process of history, diversity of everyday life becomes complicated and overtakes concepts of higher purpose, which serve as the foundation for inspiring synthesis. Lagging synthesis leads to discord in the soul—concepts of higher purpose do not correspond to the everyday way of life, to the variety of concepts and emotions, this discord leads to the decline of culture. So scientific thinking destroys ancient religious synthesis. Overcoming crises and continuing the cultural process demand new concepts of higher purpose, new synthesis, corresponding to a new level of the differentiation of psyche.

With increasing differentiation, synthesis requires ever-increasing efforts from an individual human being. Balancing these two aspects of consciousness is difficult and is achieved through the understanding of the purpose of life; Jung (1921) called this the highest aim of every human life. Similar was Schopenhauer's idea of individuation. Even more radical was Kant, who wrote that consciousness of the purposiveness coincides with the Christian ideal of sainthood. (Let me repeat this! Not helping others or sacrificing one's life for religious ideas, nor chaste life, or upholding social values—these are just stepping stones—*sainthood is consciousness of one's purposiveness*). Consciousness and culture are developed on the edge of differentiation and synthesis. Too strong a synthesis fuses the conscious and the unconscious together into a fuzzy undividedness, the need and ability for the new disappears, as in prehistoric consciousness. The prevalence of synthesis is characteristic of Eastern cultures, with their striving for peace of soul. One payoff for achieving this peace of soul is millennia of cultural immobility. The prevalence of differentiation is characteristic of Western cultures. With differentiation overtaking synthesis, the meaning of life disappears and creative potential is lost in senselessness.

Let us explore the role of music in this complex process of "keeping the scissor blades together." Jaynes (1976) analyzed the evolution of consciousness during the last 11,000 years. Weiss and Taruskin (1984) analyzed the evolution of musical styles using available data during the last 3000 years. These two sets of changes in consciousness and in music were aligned by Perlovsky. Also, Jaynes' analysis was extended by adding the idea of synthesis. This addition was essential for understanding changes in consciousness during the last 3000 years related to the rise of monotheistic religions. This alignment demonstrated, on the one hand, that during states of strong synthesis advances in consciousness were driven by differentiation, with music differentiating the "lower" emotions, and second, that differentiation violated synthesis. To restore synthesis, music differentiated the emotions of the "highest." These emotions led to understanding the violation of synthesis by bringing it from the unconscious into consciousness. The conscious understanding helped to cope with violated synthesis and to continue the process of the conceptual differentiation of consciousness and cultural evolution. From this continuous millennial process, I have selected several examples illustrating that every step in conceptual differentiation has been paralleled by powerful advances in music; on the one hand, by bringing a new level of emotional differentiation to everyday life, and on the other, through the emotional differentiation of the "highest," which helped restore synthesis.

ROLE OF MUSIC IN CULTURAL EVOLUTION FROM KING DAVID TO "KNOW THYSELF"

Contemporary Western music originated from church and synagogal singing; according to Weiss and Taruskin (1984) "Psalmody (the singing of psalms) is surely the oldest continuous musical tradition in Western civilization." But the first Biblical reference to music at King David's time (3000 years ago) refers to "the clangourous noise of instruments ... reminds the modern reader of no Western form of divine service ... (similarly does a scene) of David dancing before the arc of God." Why? Can the theory of the role of music in evolution of consciousness discussed in this book explain this drastic difference? Possibly because there were no irresolvable contradictions in the souls of David and his contemporaries, the monotheistic idea was a sufficient basis for synthesis (Fig. 7.1). Human imperfections were sins, for which one had to be accountable before God, but the notions of sin, freedom, and personal responsibility were not yet sufficiently differentiated to precipitate existential crises. This relatively undifferentiated type of consciousness we see in the book of the prophet Amos written in the 8th century BCE, 250 years after David. Consciousness presented in this book was characterized as follows: "In Amos there are no words for the mind or think or feel or understand or anything similar whatsoever; Amos never ponders anything in his heart. In the few times he refers to himself, he is abrupt and informative..."; his speech, voice, words, emotional and conceptual contents were fused, there were no deliberations, no arguments, no choices to be made. In this period of vague consciousness, music of the divine service, like all creative forces, was directed at differentiation.

However a new type of consciousness was already rising; consciousness with self-reflection and internal contradictions. Although the prophecy of Isaiah took place only one generation after Amos, Isaiah's consciousness was ahead of his contemporaries. The impending catastrophe that he foresaw created tensions in his soul between the conscious and unconscious. This tension appeared in his visions as the antiphony of the voices of Seraphims. The first time the principle of antiphony was mentioned in the Bible, the split choirs answering back and forth, which was to become the foundation of psalmody in Jewish and Christian divine service: "Seraphim... one cried to another, and said, holy, holy, holy is the Lord of hosts." "The words sung by the Seraphim entered the Jewish liturgy... and were later adopted by the Christian church..." (Fig. 7.1).

The development of consciousness in Ancient Greece, Israel, and China remarkably coincides. In the 6th century BCE the first Greek philosopher Thales repudiated myths, demanded conscious thinking, and pronounced the famous "know thyself." In Israel, the Prophet Zechariah forbade prophecy, an outdated and already dangerous form of thinking; he demanded conscious thinking. Confucius in the 5th century BCE wrote "when we see men of a contrary character, we should turn inward and examine ourselves," and his contemporary Lao-tzu, "it is wisdom to know others; it is enlightenment to know one's self." Conscious thinking created a discord between the personal and the unconscious-universal, leading to a feeling of separateness from the world; tensions appeared in the psyche, which were mirrored in antiphonal singing. Forms of music appeared, which corresponded to the emerging forms of consciousness.

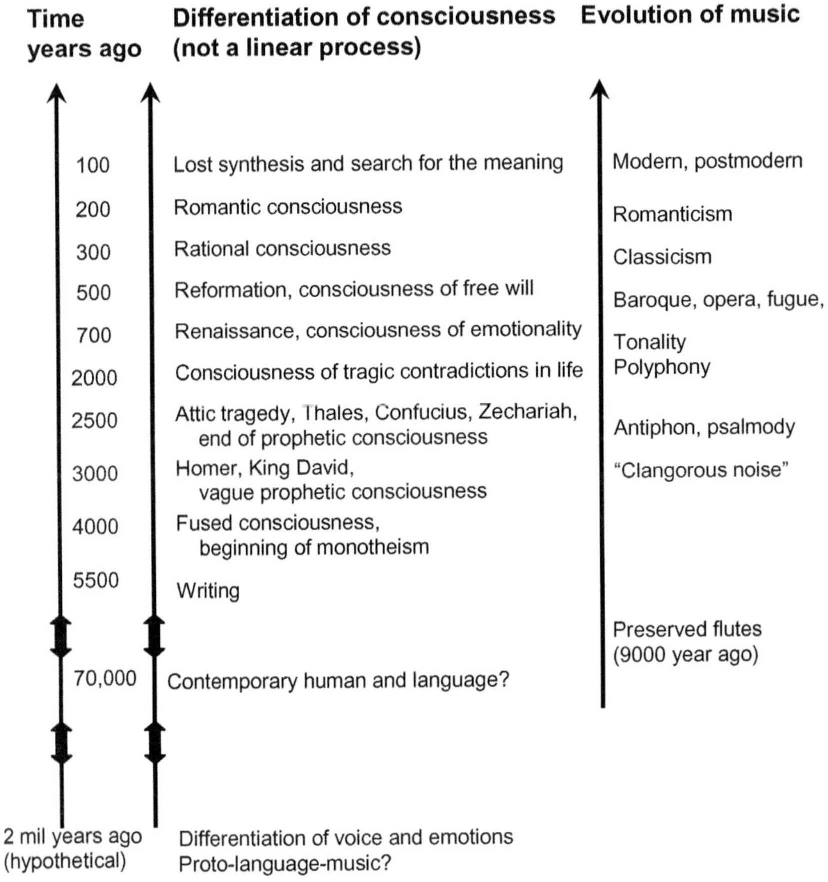

Figure 7.1 Parallel evolution of consciousness, culture, and music. Consciousness is continuously changing (as attested in language), and it requires parallel changes in musical styles for balancing differentiation and synthesis. Differentiation of consciousness is not a linear process. It can only proceed jointly with synthesis. Characterizing centennial types of consciousness with few words is a great simplification; more details are in the text and references (differentiation of consciousness from 11,000 to 2300 years ago in part follows Jaynes, 1976). Synthesis requires evolution of music parallel with evolution of consciousness; again, characterizing centennial types of music with few words is a great simplification; more details are in the text and references, in part following Weiss and Taruskin (1984).

Split-choir singing symbolized the differentiated nature of the highest principles and brought closer to consciousness the feeling of the split in the psyche. Antiphonal singing, appealing to the conscious and to the unconscious, drew them closer, linked the feeling of the split with the conscious perception of "self-world" relationships, and restored synthesis.

Antiphon as a generally accepted form of divine service is mentioned in the Bible for the first time in the book of Nehemiah in 445 BCE, just a century after Zechariah, and Thales' "know thyself."

MUSIC AND CONSCIOUSNESS IN ANCIENT GREECE

Let us consider the same period of history in Ancient Greece. Music's role in the synthesis of psyche was analyzed psychologically and aesthetically in Nietzsche's critique of Ancient Greek art. We will see amazing parallels between development of music and consciousness in Ancient Greece and Israel.

Consciousness as seen in Greek myths and legends was barely emerging from a vague and chaotic undifferentiated state, similar to consciousness of Amos considered few pages above. In Homer's poems human consciousness enjoys the force and power of language concepts overcoming chaos of ancient fused consciousness. However, sometimes, life's torrents of horror make differentiated consciousness unbearable. Man is ready to give up his individuality, forget it, and to accede to the call of unity with nature. Nietzsche's ideas of the Apollonian and Dionysian clarify the psychological meaning of differentiation and synthesis, which follow from the KI.

Life demands equilibrium between synthesis and differentiation, between an ideal meaning and a material reality. In vague mythic consciousness a human being is unified with nature and the meaning of existence is perceived instinctually, but creative ability sleeps; freedom of self-determined thinking is not yet. Differentiated consciousness opens the possibility of freedom, man begins thinking, perceives the world in clear categories of language. But at the same time, differentiation destroys synthesis, man splits from the world, and the meaning of existence becomes uncertain. When split from instincts, clear thinking may lead to lost meaning and creativity. Although in ancient Israel, synthesis was maintained with the help of monotheism, in Greece reconciliation of cognition and life was achieved through tragic drama, "salvation comes from the satyr chorus of the dithyramb." Dithyramb is a frenzied and impassioned song, ecstatic poem made up of energetic sharp lines at the verge of gaiety, sorrow, or distress. (This description equally fits a contemporary rock-concert or rap music.) In the chorus of tragedy Ancient Greek spectators met with themselves—similarly young people today are searching for their self, combining their conceptual and emotional, being lost among the complexities of life's choices.

A new form of consciousness emerged in the art of Greek tragedy, a consciousness of split, of the tragedy of existence. Tragedy emerged from the poetry of Archilochus (680–645 BCE), who introduced folk song into art. In a song, Schopenhauer saw alternations of unsatisfied willing and pure contemplation, as if musical cognition. Nietzsche modified and deepened this thought, bringing it close to mechanisms of the KI—music connects human conscious and unconscious. Music alone explains to us how it is possible to experience delight in tragedy.

"This difficult proto-phenomenon of Dionysian art is directly understood in the surprising existence of musical dissonance; in music viewed alongside with the world... If we could think of dissonance in human image... (then we understood that) existence and world are justified only as an aesthetic phenomenon. The delight created by tragic myth has the same origin as the delight in the dissonance in music... For now we can really grasp what does it mean to desire contemplation and still to go beyond contemplation. The auditory analogue of this experience is musical dissonance... a need to hear and at the same time to go beyond hearing. This striving for Eternal, sweep of the wings of yearning..."

By combining words and music, tragic drama speaks to a man as a whole, to his concepts and emotions, and mends the edges of split conscious and unconscious. Tragedy was transformed into art; this stopped chaotic swings of psyche between divided concepts and emotions. A symbol of tragedy of human existence breaks into consciousness, accelerating the process of individuation. In essence, "all the famous characters of the Greek stage, Prometheus, Oedipus, etc., are only masks of" Dionysus. Hundreds of dramatic tragedies written by great Greek authors explore one universal tragedy of individuation—emergence of Apollonian consciousness and its separation from Dionysian unconscious. And the tool used by art to penetrate into this mystery is music combined with poetry. In this combination music overcomes sufferings of individuation. But in parallel, differentiation accelerates and destroys the tragic symbol. Similarly, Protestant consciousness striving for individuation eventually destroyed the ancient symbol, and similarly during 70 years of the Soviet state, the power of communistic symbol was destroyed.

But return to Ancient Greece. In Aeschylus' dramas (524–456 BCE), tragic effect is achieved by combining emotionality of music with conceptuality of poetry. The aesthetic means are dithyrambs performed by the chorus of satyrs, animals–humans; this emphasized the fused animalistic state of psyche. Gods and mythic heroes act on a stage, and the basis for events is the archetypal myth. Aeschylus' student Sophocles (497–406 BCE) also uses mythic plots but he reduces significance of the chorus and amplifies poetic text. This emphasizes conceptual content and separates it from music. Thoughts are split from unconscious, individual from collective. And already Euripides (480–406 BCE) transported onto the stage the common man, people with their own thoughts. (This was a revolution in the history of thinking.) From the entrails of collective consciousness that thinks by clear images of objectified myths, a new type of consciousness emerges, unprecedented, unknown in history—a man is predestined to decide on his own what and how to be. Inconceivable power needed for this revolution Euripides drew from the philosophy of Socrates (470–399).

Socrates called to life a new type of rational consciousness, which creates representations in the mind according to its own understanding, without support of collective myths. Armed with the idea of individuality, rational consciousness captured the world and formed a foundation for the scientific method of thinking, which fruits are used by all of us. However, as dynamic logic explains, rational method is differentiation split from the emotional depths of unconscious. The chasm between new consciousness and unconscious was too deep for Greek culture. Ancient myth turned into tall-tales that did not grab by the guts, a symbol turned into a sign. Logical argumentative consciousness that splits from instinctive bases of psyche cannot provide a foundation for life energy. Each argument finds its counter-argument and there is no inspiring force to defend one's spiritual values in the severity of the surrounding world. Logical arguments were not sufficient for an Ancient Greek to defend his home, his country, his values. New Socratic way of thinking destroyed a bridge "over the abyss of soul." Greek culture died, writes Nietzsche, and the entire ancient world died with it. Ancient Greek civilization was succeeded by new peoples, whose consciousness was less differentiated, but intimately connected to unconscious, to foundations of psyche and to the source of living energy. Only two centuries passed from emerging individual consciousness of Thales' "know thyself" (624–546 BCE); in these two centuries consciousness and music passed through differentiation of self toward consciousness of tragedy of human existence, synthesis in the unity of language and music in Attic tragedy, and destruction of synthesis in Socratic logical thinking.

Recapitulating, I would like to stress that evolution of Ancient Greek consciousness is but one period in history of complex dynamics of differentiation and synthesis. Two periods of Ancient Greek flourishing,

Homeric and tragic, are times when conceptual and emotional were unified. During the period of Homer, ancient synthetic consciousness gained access to clear concepts (differentiation) through language enriched by writing; possibly for Homeric poets poetry was unified with music. During the period of Attic tragedy, torn differentiated consciousness was unified in a symbol of tragedy—with the help of music the symbol unified conscious and unconscious. Rational Socratic consciousness joined with Euripides however, uncovered the irrational nature of myths and gods, the unconscious roots of musical tragedy. The idea of tragedy was differentiated. Synthesis was destroyed.

TONAL ORGANIZATION SINCE 250,000 BCE

Ever-increasing complexity of human cognitive world required the parallel increase of complexity of music. How far back can this process be traced? Aleksey Nikolsky uses the data of comparative ethnomusicology to identify variety of methods of tonal organization of music before 250,000 BCE. Here I follow Nikolsky in using "tonal organization" for every form of integrating musical tones in a harmonious ensemble. The strategy employed for this purpose by a particular culture reflects a general cognitive approach to orientation in the world employed by the members of that cultural group.

Nikolsky establishes 14 discrete methods of tonal organization, which he traces to before 250,000 BCE. He uses the findings of developmental psychology of music perception to align these 14 methods in a chronological order. The modern newborn, he suggests, is essentially in the same state of musical skills as the Neanderthal newborn—except that the Neanderthal baby did not have models of tonal music performed in his environment. Therefore the kind of melodies that modern babies spontaneously vocalize at around 2 years of age must essentially be the same as what was used by *Homo heidelbergensis* until Neanderthals, Denisovans, and *Homo sapiens*.

Such melodies are characterized by free gliding, where only the direction of the melody counts, and all other parameters are random. The next transition occurs at about 250,000 BCE, when the first colored or marked stones were created. The skill to mark a rock is likely to correspond to the skill to mark vocalization by unusual twist in the vocal timbre and pitch. The next method of tonal organization emerges—where the singer joins the musical tones not by setting the sequence of directions for smooth gliding motion, but by contrasting smooth gliding motion to abrupt jumps from one vocal register to another, accompanied by timbral transformations. This method corresponds to children's babbling at the age of 2–3 years. Other stages similarly build musical skills and cognitive schemes, one on top of another, together, defining the line of development of greater sophistication of tonal organization until the formation of modern tonality.

What is interesting is that the entire progression of stages demonstrates not only increasing complexity in hierarchic subordination of musical tones but also increasing the amount of pitch units employed to build musical composition, and progressive narrowing of the normative pitch values for each of the pitch units. A 250,000 BCE musician found his way in music-making by distinguishing between two musical registers, with each up to an octave wide (12 semitones). A modern Western musician distinguishes among 17 different pitch-classes, each of which is tuned differently, and the width of the tuning zone that is conventionally "right" is about a half of a semitone.

Such progressive complexity and concision remarkably coincide with Kapitsa's depiction of the evolution of human knowledge. Nikolsky also identifies similarities in development between tonal organization in music and spatial organization in pictorial arts. It seems that indeed human mind began its cognitive growth during Paleolithic times, and went through stages of cognitive development with ever-increasing pace.

This pattern of development is a manifestation of ever-increasing pressure of growing cognitive dissonances, and the growing burden for the music system to keep them in balance by offering greater and greater layers of harmonization through complex hierarchic coordination of musical sounds.

SYNTHESIS AND DIFFERENTIATION DURING EARLY CHRISTIANITY AND MIDDLE AGES

At the beginning of the Christian era, anxiety often did not find justification or an outlet. Similar feelings are often experienced by contemporary people, but 2000 years ago the symbol of God suffering on the cross created a mystical sensation of meaning and pacified contradictions in the soul. Such a mystical and powerful process in psyche Jung called the autonomy of a symbol-process. When a symbol acts autonomously, it assimilates a larger part of sufferings and contradictions. Remaining contradictions were reconciled in the synthesis of antiphonal music.

Collective emotions of the spiritually sublime expressed in antiphonal singing were adequate for synthesis.

"The early Christian church adopted its rituals... from the liturgy of the synagogues... Jewish liturgy... became the framework for the early Christian vigils... Modern research has uncovered a close resemblance between certain psalm melodies preserved to this day among Middle Eastern Jews and certain Gregorian chants... The singing is antiphonal... with refrains, which in Christian times became known as antiphons."

Descriptions of music as a mechanism of synthesis, unifying meanings of words with unconscious, we encounter in the church fathers in the 4th century: "The delight of melody He mingled with the doctrines so that by the pleasantness and softness of the sound... we might receive without perceiving it the benefits of the words." (The meaning of antiphonal psalmody was identified with a significant part of Christian teachings and the purpose of music was seen in connecting everyday life concepts with the sublime, i.e., in synthesis.)

Collective experience of spiritual unification, which we called synthesis, was adequately created and expressed by existing musical forms, monody and antiphony. This was described in the writings of St. Basil and St. Augustine: "(music was) joining the worshipers into a unanimous entity", "in the practice of singing hymns... the faithful fervently united with heart and voice..." Synthetic power of music was also discussed by Boethius: "...what unites the incorporeal existence of reason with the body except a certain harmony, and, as it were, a careful tuning of low and high pitches in such a way that they produce one consonance?"

However, as the KI asserts, history is defined by complex dynamics of synthesis and differentiation. Inspiration created by early Christian synthesis released creative power for differentiation of musical forms and accompanied mental awakening. This shifted balance in the soul and threatened to destroy synthesis. Development of consciousness in Christian Europe reminds processes in polytheistic Ancient Greece. Again, powers of music were recognized in their complexity, constructive and destructive, similarly to Plato's view of poets and musicians as a threat to culture. Music appealing to regressive instincts is repudiated by the church as a destructive force: "Passions sprung from a lack of breeding and baseness are naturally engendered by licentious songs... at the sound of a flute (some) are excited to a Bacchic frenzy." This ambivalence toward music was due to music's real powers of influencing people, hence the suspicion of strongly emotional music. (A similar attitude we can see in our time, the so-called serious academic music in the 20th century have reclused itself from the world of emotions, strong experiences, and nontrivial meanings.)

Ambivalence toward music is a result of more complicated dynamic interaction between differentiation and synthesis. We find its self-reflective depths in the 4th century in St. Augustine: "The tears flowed from me when I heard your hymns and canticles, for the sweet singing of your Church moved me deeply. (When singing) the jubilus... it is a certain sound of joy without words, the expression of a mind poured forth in joy... There are particular modes in song and in the voice, corresponding to my various emotions and able to stimulate them because of some mysterious relationship between the two. But I ought not to allow my mind to be paralyzed by the gratification of my senses, which often leads it astray... Senses... attempt to take precedence and forge ahead of (the mind)..."

St. Augustine was afraid to lose control of mind over emotions. According to foremost thinkers in the 4th and 5th centuries, the mind was not strong enough to be reliably in charge of senses and unconscious urges. Differentiation of emotions and feelings was perceived as a danger. Boethius, 5th century, expressed a firm belief that in an individual human being the highest was associated with conceptual reason that ought to judge sensual and emotional: A musician is the one who maintains "the sovereignty of reason ... and not through the slavery of labor" learned music. Instrument players and poets (composers), according to Boethius, are not musicians, because players "devote their total effort to... skill... use no reason, but are totally lacking in thought," and "compose songs not so much by thought and reason as by a certain natural instinct." Differentiation of emotions in the 4th century was dangerous for synthesis, which was not yet ready to unify the entire material sphere of human existence. Therefore early Christian music was directed at spiritually sublime emotions with synthesizing power. In religious and social life, similarly, unity in soul was fragile and creativity was directed at strengthening synthesis at the price of repudiating the material and innovative.

Today, attitude toward musical emotions is ambivalent as well. On one hand, apprehension of overly emotional music can be observed in minimalism, in attempts to create "intellectual" music devoid of emotions and to separate "serious-conceptual" music from "popular-emotional." On the other hand, huge crowds in concert halls and arenas seemingly lose control of the mind. Is there a threat today to culture from "wrong music," or has consciousness become strong enough to rebuff such a threat?

In the 4th century Christianity was not ready for synthesis of material and spiritual. Repudiation of material life contradicted foundations of the Roman Empire. When Christian consciousness penetrated into souls of the majority of Roman citizens, the empire disintegrated. In the 5th century Western Europe was subjected to a series of invasions by Visigoths and Huns. The Roman Empire's systems of agriculture, roads, water supply, and shipping routes gradually decayed, and the unity of imperial society, government, and culture was replaced by the conflicting powers of the various Germanic tribes. Artistic and scholarly work also faded. It took four centuries to create foundations in the collective soul for unifying spiritual and material. Christian states appeared in the 8th century, and much of Europe was unified under Charlemagne, who codified the laws and promoted a cultural revival. Cities, population and prosperity grew, universities opened, philosophical thought was reacquainted with Aristotle.

Synthesis in psyche was fortified about the 8th century, when Christian states emerged. Creative powers could again be directed at differentiation of consciousness. Unification of opposites, a conflict between rational thought and mystical foundation of religion created tensions in the human soul. The emerging tension inspired creativity in search of synthesis. New musical forms of sequences and tropes unified conscious and unconscious in the interaction of conceptual contents of texts and emotional contents of melodies. A popular song today affects us in a similar way.

An important step toward differentiation of musical content was the invention of musical notations in the 9th century. This was a revolution in the history of Western music determining "all future development of the art... musical style began to change with increasing rapidity... (It) made the cultivation of polyphony on a grand scale really practicable... which above all sets Western music apart from all other traditions." From this time on originality and individuality of music is valued, composers' status is elevated, differentiation of internal experiences, feelings and emotions is accelerated. In an early form of polyphony, organum (9th century), two voices move in parallel motion. Organum "reminded of the universe (that) in accordance with the uniform will of the creator, is welded into one harmonious whole..."—this witness of Erigena (815–877) confirms the previous scientific conclusion that music is understood by both conscious and unconscious and is capable of differentiation as well as synthesis.

During the following centuries, organum evolved toward richness of emotional content in greater melodic independence. Differentiation of experiences and emotions did not correspond any longer to antiquated Christian synthesis based on repudiation of worldly pleasures. Lagging synthesis again threatened a cultural catastrophe. In 1159 bishop John of Salisbury protested against danger perceived in new music: "the singers... with lewdness of a lascivious singing voice... To such an extent are the high or even the highest notes mixed together with the low or lowest ones—that... the intellect, which pleasurableness of so much sweetness has caressed insensate... Such practices... can more easily occasion titillation between the legs than a sense of devotion in the brain." In the bishop's mind, clearly the lower emotional appeal of this music was stronger than its synthetic power to combine the material with spiritual.

It is interesting to note that restraining of musical emotions today comes not from the church but from antiemotional movement in academic music, such as minimalist' music by Cage (1912–92) or Glass (1937–), or serial music by Boulez (1925–2016), to name just most famous ones. I think today this direction toward impoverishing musical emotions is related in part to misunderstanding of the role of music in consciousness, cognition, and culture. Connecting music to mathematics by some composers is a spiritually similar idea. But return to the High Middle Ages music of courtly love.

Romantic idea of love emerged in worldly life, feelings to a beloved one took on a religious tinge in the 11th–13th centuries. Variety and refinement of emotions began to flourish in courtly life and the knightly tradition. In secular songs of troubadours and trouveres in France, minnesingers and meistersingers in Germany, emotions are differentiated more than ever before. "Like the music of the Church, the poetry and music of the noble troubadour were sober and reflective, and served to elevate and memorialize the permanent values of life... service to lord and lady, the idealization of love, and the fervor of Crusades." The idea of romantic love was elevated to a level of the highest spiritual purpose and love feelings were elevated to the level of religious ones. Differentiation and synthesis of the highest and lowest emphasized in collective consciousness the most powerful instincts, material and spiritual, an opposition previously suppressed. Romantic sublimation of love became a part of the new synthesis of emotional and conceptual—it was a step toward individuation becoming such an important element of Western culture. Poet-composer in a song turned to his beloved one—an "eternal theme" of love celebrated yet in Solomon's Song of Songs, firmly entered Western culture only about one thousand years ago.

Combination of diversity and meaning, differentiation and synthesis, was accompanied by enhanced interaction of music and text. Musical techniques of secular songs emerged, which are still used today: instrumental accompaniment, stronger and more regular rhythms, recurrent short rhythmic patterns and refrains; instead of Latin, texts in vernacular languages accessible to understanding enhanced synthesis of the secular and mystically high. In secular life, differentiation and synthesis got ahead of these processes in the life of Church—the Christian symbol, having created during the millennium a new type of consciousness with a higher level of individuation, began losing autonomy in the human soul, its role as the spiritual source of life was waning.

INDIVIDUATION: THE RENAISSANCE, REFORMATION, AND BACH

During the Renaissance (the 13th–14th century), for the first time since antiquity a European man felt the power of a rational mind separating from collective consciousness. A new consciousness engages and inspires people only if it creates a new synthesis in place of the old decaying one. A process of creating synthesis, new symbols, inevitably goes through tragedies and catastrophes of the demise of the old ones. Human intellect is not as omnipotent as it may seem to collective consciousness. However, in the beginning of the Renaissance synthesis was strong, backed up by both, a new symbol of the greatness of human reason and by ancient religious mystical symbols; the result was a creative explosion. Polyphony, "probably the greatest... development in the entire history of Western music," defined the next step in evolution of consciousness, where the idea of the individual as a part of the highest began to permeate into collective consciousness.

The principal polyphonic form during the 13th century became motet; it was more complex than organum, in several different parts with independent melodies, texts, and rhythms sung simultaneously, and "everywhere sounding in harmony." But already, differentiation was overtaking, the unity of concepts and emotions became of less concern: "the words are fitted to the music only after the music is completely composed, and then only 'as best one can'." The millennial tradition of music perception was changing. For twelve centuries, Plato, Boethius, and

Erigena (from the 4th century BCE to the AD 9th century) saw the positive content of music in its relations to objective "motion of celestial spheres" and to God-created laws of nature. This changed by the 13th century: The music was now related to listeners, not to celestial spheres. Songs for "average people... relate the deeds of heroes... the life and martyrdom of various saints, the battles..."; songs for kings and princes "move their souls to audacity and bravery, magnanimity and liberality... about delightful and serious subjects, such as friendship and charity... a motet ought not to be propagated among the vulgar, since they do not understand its subtlety... but it should be performed for the learned... for their edification."

From the "objective," music moved toward human feelings. In the 14th century the first musical avant-garde emerged; Ars Nova (The New Art) used notes of variable durations for differentiation of emotions. Change of musical technique paralleled differentiation of European consciousness: mystical power of the religion diminished, diverse emotions created in music "distracted" from religious life. In 1323 Pope John XXII criticized the new music (Weiss and Taruskin, 1984, p. 71).

"By... dividing of beats... the music of the Divine Office is disturbed with these notes of quick duration. Moreover, they hinder the melody with hockets (interruptions), they deprave it with discants (high-voice ornamental melodies), and... pad out the music with upper parts made out of secular songs... The voices incessantly rock to and fro, intoxicating rather than soothing... devotion... is neglected, and wantonness... increases. Nevertheless, it is not our wish to forbid the occasional use of... (polyphony), which heighten the beauty of the melody... (Polyphony) would, more than any other music is able to do, both soothe hearer and inspire his devotion, without destroying religious feeling in the minds of the singers."

The Pope as if foresaw the crisis of culture due to the lost beliefs. The Christian symbol was losing autonomous power in the human soul. Trouble was all over Europe, strifes among nations and social classes, Papal exile, schism within the Church, The Great Plague. The catastrophe coincided with a lost unity within the soul, and again, chasing the lost wholeness, a cycle of the restoration of synthesis wound up.

What kind of music could inspire people, when the power of the mysterious was lost and the dominating idea was humanism, the power of human reason? Reason accelerated differentiation of concepts, including the highest concepts of meaning and purpose. According to the KI mechanisms, synthesis requires a balance between concepts and emotions. To restore synthesis, more diverse emotions were needed. Beginning in the Renaissance and until our time, a musical system of *tonality* was developed for differentiation of emotions.

Tonality is the system of functional harmonic relations, governing most of Western music. Tonal music is organized around *tonic*, a privileged key to which melody returns. Melody leads harmony, and harmony in turn leads melody; a melodic line feels closed, when it comes to rest on (resolved in) tonic. Emotional tension ends and a relaxation is felt in the final move on to the tonic, to a resolution in a "cadence." The fundamental role of tonality in Western music for over 500 years points toward the a priori, inborn mechanisms of the mind related to the previous discussions of musical perception. However, scientific studies of these mechanisms are not yet sufficiently advanced, we know how the ear perceives consonances and dissonances, but not complex tonality. It is not clear which mechanisms of tonality perception are inborn and which are culturally transmitted.

Music connecting differentiated emotions with the sublime emerged in the 15th century. John Dunstable, according to contemporary witnesses, changed all "music high and music low," music became more consonant and euphonious. Melody and rhythm were concentrated in the top part, supported by chordal harmonies. Franco-Flemish or Netherlands school emerged, including Guillaume Dufay, Johannes Ockeghem, Antoine Busnois. "Harmonies exalted even heaven... like angelic and divine melodies... (As if the) songs of angels and of divine Paradise had been sent forth from the heavens to whisper in our ears an unbelievable celestial sweetness" (this refers to 1436 performance, "most likely to the Dufay's grand motet 'Nuper rosarum flores'").

New music created inspiration synthesis that freed creative forces for differentiation and took over multiple aspects of cultural life. Gutenberg printed the Bible—spiritual truths could be investigated by an individual human being outside of the collective church interpretation. A corporate democracy emerged in Holland. Petrucci printed polyphonic music. (Music was no more an exclusive property of churches and castles, but it was for the *public buying*, from then on the composer's income and therefore music would be affected by public tastes.)

The Renaissance synthesis was based on humanism, human values: "music's true purpose and content... (is its) power to move" emotions. This thought "a medieval thinker would have found incomprehensible... The new Renaissance attitude... valued the natural, spontaneous gift of the artist over the application of reason and mastery of theoretical doctrine." Attitude to emotions in music was changing. In the beginning of the Christian era St. Augustine was afraid of unruly emotions, balance in soul and synthesis required repudiation of material and suppression of passions. Now, after 1500 years of Christianity, man was becoming (to some extent) a master of the self. Untamed emotions were no longer considered a morbid threat to society, self, and spiritual interests. Humanistic ideal had inspired the Renaissance man to look for increasingly stronger emotions—and this search continues today (although an opposite tendency, fear of strong emotions, has reappeared in the 20th century Academic music). Polyphony became complicated, harmonic orientation enhanced, correspondence of text and sound strengthened—from Dunstable to Desprez differentiation (in texts) and synthesis (in music) were developed together. Variety of human emotions was successfully combined with spiritual aspirations.

Religious polyphony of the 16th century found its highest expression in major scales and angel choirs of Palestrina. Heavenly harmony in Palestrina's music represented one side of the soul of a medieval Christianity, absolute purity expressed in the idea of Christ personality. Christianity raised the ideal into the heights unreachable to man, but this also delineated, emphasized existence of evil as absolutely opposite to the Divine. Absolutization of evil was confirmed by the practice of inquisition, in essence acknowledging evil as a self-governing force. However the idea of a real existence of *evil* long remained unconscious. Medieval theological concept of evil as *privatio bono (absence of good)* meant that evil did not exist. Similarly, in musical tradition from early Church psalmody to Palestrina, dissonant chords were not encouraged, tritone chords sounding with complex emotions of major and minor at once were forbidden—as if evil did not exist. Thought endured a contradiction: in theology and music evil did not exist, but evil existed in life. Unconscious (undifferentiated) status of evil was an obstacle for the process of individuation.

The highest ideal of Christianity, improvement of inner spiritual life, traditionally demanded repudiation of the material world perceived as a temptation and distraction from the highest spiritual purpose. The best way to achieve the ideal of sainthood was supposed to be a monastic way of life and a rejection of secular life. However, rejection of the world acknowledged the absolute power of evil within the material world. By the 15th century the ascetic ideal came to contradictions with developing rational thinking and the emerging capitalistic economy. The *Reformation* in the 16th century accepted that the highest human calling was in perfecting the inner spiritual world as well as the outer material world (and the material conditions of one's life). The religious ideal was reconciled with this new consciousness.

Highest human purposes Luther brought closer to everyday rational pursuits. The church was to become the community of believers and religious life more rational. Instead of the monastic choir, the entire congregation shared in church singing. This was the purpose of the Lutheran chorale, the unaccompanied congregational hymn, sung in unison and in the vernacular (Weiss and Taruskin, 1984, p. 103).

Note that opera, discussed later, often is performed in the original language, which many listeners do not understand. The reasons are many. Fashions stimulate singers to pretend inculturated, multilingual. But an opera star traveling around Europe cannot learn classic operas in many languages. This sacrifices the synthesis of text and music to meaningless fashions. Musicals and operettas are performed almost exclusively in the language of the listeners.

Luther saw in music the synthetic power that unifies the Word of God with human passions: "Therefore... message and music join to move the listener's soul... The gift of language combined with the gift of song was only given to man... (so that He proclaims) the word through music."

The Reformation reduced the absoluteness of the split between spiritual and material, good and evil—the contradiction between good and evil was taken from the heights of Heaven and the depths of Hades and placed into the human soul. From then on, a protestant woman or man was to decide on his own, within his own consciousness how to reconcile contradictions between material and spiritual needs. Consequences were on the one hand an inconceivable acceleration of the development of capitalism and improvement of material conditions of life. On the other hand, the autonomy of religious symbols was lost; their unconscious contents were to a large extent transferred into consciousness. The fundamental contradiction of human nature between finite matter and infinite spirit, which formed the mystical foundation of Christianity, was brought by The Reformation into everyday culture and made a part of collective consciousness. Never before had collective consciousness placed such a degree of responsibility on a human being—for one's own life in this world and for one's own eternal soul. Tragic tensions originally projected onto the Christian symbol were assimilated by human psyche. *Tensions in the human soul reached the maximum.*

The naïve humanism of the 15th century barely glimpsed into the contradictions of human thoughts. Consciousness of the mind's internal contradictions was an achievement of the Reformation, and this consciousness required new forms of synthesis to restore wholeness and to continue evolution of culture. In search of synthetic forms of art creative minds turned to the epoch of crises passed, when salvation was found in art. From the books of Plato, Aristotle, and other authors of antiquity, it was known that tragic musical drama in Ancient Greece created catharsis, an intimate bond with the human soul, which miraculously calmed discontent, soothed character and behavior. "Radical humanists" in the 16th century sought to recover the true music of antiquity, which according to their ideas was in close connection with rhetoric, the art of orators and actors. A literary expression of these ideas was given by Vincento Galilei (Florence 1588, he was the father of the great astronomer). A new form of music, "musical speech," or recitative, quickly led to the true opera "Orpheus" by Claudio Monteverdi (1600) and made a profound influence on the subsequent development of Baroque music.

The Baroque was full of dualism and drama, expressing tensions imposed by the Reformation.(It is a world searching for differentiated synthesis.) Dualism was embodied in a new musical style, where oppositions were emphasized: Vocal against instrumental, solo against ensemble, melody against bass, dynamic levels were contrasted, opposition of the dominant and tonic expressed emotional tension and resolution. The role of dissonances increased, and modulations became commonplace (transitions between tonalities) expressing more and more complex emotions in their continuous flow. Creating emotions was becoming the primary aim of music; composers strived to imitate speech, the embodiment of the passions of the soul. At the same time conceptual content of texts increased, "the words (are to be) the mistress of the harmony and not its servant" wrote Monteverdi. Thus, *conscious aims of Baroque music were differentiation of emotions in parallel with synthesis of conceptual and emotional, which is, individuation of consciousness.*

But Monteverdi's idea of individuation, equal play of words and music in opera could not survive for several reasons. One of the reason is complexity of personal interactions between a composer and librettist (the author of the poetic text) considered later. It is interesting to trace transformations of the Monteverdi's dictum: "The words (are to be) the mistress of the harmony and not its servant," (1600). Gluck: "Text ought to be connected to music as music to text" (1750s). Motzart: "Words are servants for music" (1780s). Pushkin: "Even for Rossini I would not move a finger" (1810s). Music acts on emotions directly, therefore in synthetic art forms, like songs, musicals, operas, music commands the attention of listeners, even if lyric is deeper and artistically higher than music. A poet is also a dramatist and largely determines contents of a musical, but public remembers the name of the composer. Therefore many great poets did not collaborate with musicians and hated songs written on their verses, even as these popularized their names. Schostakovich wanted to write a series of operas about Russian woman characters, but he wrote just one ("Lady McBeth..."). Seven times he was ready to start and seven times he obtained funds. But successful collaborations among composers and lyricists often fall apart, because public bestowes fame on composers disproportionately. Such is a priori power of musical sounds over our emotions.

During Baroque the nature of emotions became a vital philosophical issue. Long before that Descartes attempted a scientific explanation of passions (1646). He deeply penetrated into the interaction between emotions and consciousness: "those who are the most excited by their passions are not those who knows them best and their passions are... confused and obscure." Still, he did not differentiate thoughts, actions, and emotions consistently: "of all the various kinds of *thoughts*... these *passions*," "the *action* and the passion are thus always one and the same thing." Descartes showed a bias toward unconscious and fused perception of emotions, which is characteristical of his thinking psychological type. He rationalized emotions, explaining them as objects and relating to physiological processes. "Descartes descriptions of the physiological processes that underlay and determined the passions were extremely suggestive to musicians in search of technical means for analogizing passions in tones." Based on Descartes' theory, Johann Mattheson formulated a new theory of emotions in music, called "The Doctrine of the Affections." "Now the object of musical imitation was no longer speech, the exterior manifestation of emotions, but the emotions themselves."

Already during Palestrina, young Italian composers began using dissonances to express passions in madrigals; however until the end of the 16th century dissonances were used sparingly, for a short pause, and mainly in secular music. Beginning in the 17th century, dissonances were used more often, emphasizing the dramatic effect. A dissonance was always followed by a resolution in a consonant chord; later several dissonant chords were used in a row, *increasing tension*. The heightened sense of drama in musical dissonances corresponded to the tension between conceptual and emotional, material and spiritual, which in the result of the Reformation being assimilated by human heart and soul.

Human everyday emotions had never been as diverse as expressed in opera. Personal expressions grew even in seemingly minute details as the signs "allegro, forte," which composers inscribed on their works. Music became extremely expressive, conveyed passionate human emotions; theory of harmony was developed for this purpose, chromatic scale was used (chromatic scale contains all 12 pitches separated by a semitone; in tonal music it is used as decorative runs up or down as it has no harmonic direction; it might be considered a step toward 12-tone music). Chorale was unified with counterpoint, harmony with polyphony. These new musical forms were perfected in works of Buxtehude and then Bach. The most complex and sublime form of polyphonic music acquired in fugue: polyphony is combined with harmony, in the "horizontal" of melodic and "vertical" of harmonic spaces of sounds. Fugue is a conversation of several musical voices, in which a topic "flies" from one voice to another; voices could talk politely or argue, interrupting each other. In Bach fugues a human arguing with oneself turns to God or to the highest in oneself.

Although old psalms affirmed an existence of the objectively sublime, as some collective purpose far removed from individual experiences, fugue expressed emotions of one's own contradictions in quest for the highest. Fugue was a way of individual consciousness turned to sublime, a culmination of a millennial process of individuation, a combination of differentiation and synthesis. Rational understanding of Church service introduced by the Reformation interacted in music with the highest spiritual values and mystical feelings of sublime, created during thousands of years by monotheistic religions.

Yet, rational consciousness that came after the Baroque rejected mystery of sublime differentiated in fugue. Music that was natural to Bach seemed too intellectual and "not natural" to the next generation. We know that when spontaneously improvizing, Bach excelled especially at those aspects of his music that seemed too intellectual to the next generation.

The Reformation has laid unbearable responsibility on an individual and created too much tension within the human soul—humankind is not ready yet for individual consciousness. The string of tension connecting conscious and unconscious broke.

CLASSICISM AND RATIONALITY

A new style Rococo appeared in French clavecin school in Francois Couperin, although Bach and Handel were still writing music. "The heroic dimensions of the Baroque were cut down to a more human scale... a graceful decorativeness and sweet intimacy of expression were prized above all." In Germany, leaders of this style were Bach sons, C.P.E. Bach, J.C. Bach, and W.F. Bach; they simplified harmony and texture, abandoned counterpoint and polyphony, retained one leading voice, sometimes with a simple accompaniment. Restrained expression and aristocratic style, ornamentation, delicacy and obsession with structural clarity corresponded to secular interests that replaced a search for the highest meaning. Interests moved "away from "reason" and back to "the ear." Contemporaries were convinced that the superiority of the new music was a matter of "ear" or "good taste." But looking back, we can see that the change of musical forms corresponded to the underlying change of consciousness: An idea of the *highest* was lowered to the "human scale."

Why were contradictions of Baroque superseded by the simplicity of Rococo? Between Renaissance and Reformation, as mentioned, interaction of conscious and unconscious—synthesis based on mystery lost autonomous power. The Christian symbol did not touch unconscious archetypes as forcefully as in previous epochs, and the Reformation was a theologic and political acknowledgment of this process. After the Reformation, the creative process was inspired by reason more than by mystery. The idea of rationality was amplified by scientific method. In collective consciousness there emerged a *myth of rational mind*, which substituted ancient symbols.

The word *myth* here emphasizes a *symbolic* rather than rational nature of the idea of rationality. Many scientists appreciate limited powers of reason; Pascal wrote "the supreme function of reason is to show man that some things are beyond reason." But among lay public the idea of reason replaced the old religious symbol. As discussed later in the book, with the advent of science a new synthesis was created on the basis of this idea. In essence, rational was assigned a mystic power to resolve unresolvable contradictions in the human soul. The complex nature of the "myth of rationality" has remained unconscious. A *wrong* idea has emerged in collective consciousness that science and mind are limited to the rational and conscious, whereas irrational and unconscious are obsolete notions having no place in science and "contemporary" culture.

In this period of rationality collective consciousness of Rococo replaced "mystery of the highest" with refined, ordered, aristocratic, and rational—mysterious and unconscious were perceived as unnatural and in poor taste. The epoch of mystical contradictions was dying out; sharp contradictions in the human soul retreated into the past and were superseded by a belief in rationality of the mind. Fugue, having been created for achieving synthesis in a soul torn by contradictions, was too complex for perception; it demanded spiritual efforts which seemed unjustified, when synthesis was based on *the rational.*

The idea of rationality influenced all areas of cultural and social life; Rococo is a time of Enlightenment and rationalism in European consciousness. Rationalistic thinking drew *inspiration—i.e., synthesis*—from successes of science and Newtonian physics. The new symbol of rational scientific thinking accomplished synthesis of material and spiritual without penetrating into mysteries of the highest purposes. Mystery as if disappeared, as if there were no unsolvable contradictions between reason and passions. Displaced from consciousness, mystery "went" into rational endowed with Divine powers. The new synthesis created conditions when creative efforts could be concentrated on differentiation of the everyday experience; music concentrated on differentiation of emotions.

Rational, secular, clear ideals of Rococo were continued in the Classical period in music of Haydn, Mozart, and early Beethoven. Although Bach talked to God directly, Mozart could only appeal to the beauty of his melodies. He brought everyday people onto a scene. A similar transformation of consciousness from sublime to everyday occurred in Ancient Greece, when gods and mythic heroes of Aeschylus and Sophocles were replaced in Euripides' plays by everyday people. But whereas in Greek myths *the rational* has just begun appearing from unconscious fuzziness and their comedies are emotionally simple, in Haydn and Mozart there is a finest differentiation of emotions. European collective consciousness in the eighteenth century has been vastly different from Ancient Greece.

Transformation of consciousness toward rationality is seen in the development of opera. It became coherent work of art, in which drama, text, and plot played equal roles with music. The opera reform was began by Gluck and Calzabigi in the 1760s, and later completed by Mozart. This integration of text, drama, and music created *synthesis* of conceptual and emotional. Fine differentiation of emotions was expressed in melodic chromaticisms and modulations—deviations from exact tonality, interchanges of tonic major with minor. Haydn and Mozart used these techniques for creating spontaneous changes of emotional states. Although 2000 years ago, Euripides' revolution destroyed the tragic symbol along with the unity in the collective soul of the ancient world, Haydn and Mozart, by differentiating emotions of common men and women, participated in creating new synthesis.

Modulation is a change from one key or tonal center to another. In much of tonal music the tonal center is perceived psychologically as a "center of gravity" attracting the listener's expectations. A modulation or movement between tonal centers creates a dramatic effect, a feeling of "moving in a new direction." Bach made it basic to every composition; Haydn and Mozart expanded its dramatic effect, and the use of modulations for increased dramatization continued thereafter in Beethoven and in romantic music, until as discussed later, the notion of the tonal center was abandoned in atonal music.

"The expression of a child-like, serene mind governs Haydn's compositions. His symphonies lead us to endlessly green pastures, to a merry, colorful throng of happy people. Dancing youths and maidens are floating by; laughing children, hiding behind trees and rose bushes, throw flowers at each other. A life full of love, of bliss, like before original sin, in eternal youth; no suffering, no pain, only a sweet, melancholy longing for a figure that floats by in the distance, at dusk, and does not come nearer, does not vanish, and, as long as it is present, it does not turn into night, since it is the evening glow, itself, in which mountains and fields are steeped. Mozart leads us into the realm of spirits, but, without pain, it is more of an anticipation of the infinite. Love and melancholy sound in lovely spirit voices; night arrives in a purple glow, and with unspeakable longing, we move towards them who wave at us to join their ranks and to fly with them through the clouds in their eternal dance of the spheres. Haydn sees the human in human life... his music is more commensurable, more comprehensible to the majority. Mozart evokes the super-human, the wonderful that dwells in the innermost of spirit." E.T.A. Hoffmann.

For thousands of years, music expressed and created emotions, but only in the 18th century the idea of music as *expression* of emotions, *creating* emotions in listeners, was *brought into consciousness. This idea of emotion as expression led to understanding of music as art differentiating (creating new) emotions; it related the pleasures of music sounds to the 'meaning' of music.* The previous theory of *imitation* was chiefly associated with vocal music and with the "doctrine of the affects"; it considered music alongside all the other arts as a medium of stylized *representation* of reality. But this does not correspond to workings of the mind. The KI theory tells us that representations are the mind's model-concepts, whereas emotions are fundamentally different neural signals evaluating and relating concepts to instincts.

Let me repeat that according to the KI mechanisms, neural representations are the mind's model-concepts, whereas neural emotional signals are fundamentally different, they evaluate and relate concepts to instincts and other concepts. Music is different from other arts in that it affects emotions directly (not through concepts-representations). This clear understanding of the differences between concepts and emotions did not exist, and an idea of music as expression, differentiating (creating new) emotions, was not obvious. It was first formulated by Charles Avison (1753) and then repeated by James Beattie (1778). *This idea of emotion as expression led to understanding of music as art differentiating (creating new) emotions; it related the pleasures of music sounds to the 'meaning' of music.*

It may seem one should not expect a direct relationship between music and philosophical ideas about emotions. But let me remind, this is an essential content of this book, and we may marvel how directly these relationships turned out! Descartes understanding of *emotions as objects* and Matthesons' doctrine of the affects turned out directly related to musical practice of the opera seria (serious opera). Opera seria inherited ideas and practice of Monteverdi, but soon turned into their opposite. By the middle of the 17th century opera became stylized and rigidly regulated set of airs, expressing concrete emotions. *This emphasizes how important for music and for art in general is correct scientific understanding of the nature of emotions and beauty,* which still is missing in standard education of artists, and possibly is directly responsible for the large amount of garbage in the museums of fine arts and on the conservatories' scenes.

Previously mentioned Calzabigi-Gluck reform was directed against this regulated "serious" opera and wrong understanding of emotions. Although the clear understanding of the differences between concepts and emotions did not exist, an idea of music as expression was a step in the direction of appreciating the music power of differentiating (creating new) emotions. Thomas Twining (1789) emphasized an aspect of music, which today we would name conceptual indefiniteness: Music exists for creating emotions not semantics, musical contents cannot be adequately expressed in words and do not imitate anything specific. "The notion, that painting, poetry, and music are all Arts of Imitation, certainly tends to produce, and has produced, much confusion... and, instead of producing order and method in our ideas, produce only embarrassment and confusion."

The nature of emotions remained utterly confused: "As far as (music) effect is merely physical, and confined to the ear, it gives a simple original pleasure; it expresses nothing, it refers to nothing; it is no more imitative than... the flavor of pineapple." While directed at correcting the current musical practice, it is scientifically wrong. Pleasure from musical sounds is not physical and not confined to the ear, as many thought. (Pleasure from music is an aesthetic emotion in our mind, whereas we like the flavor of pineapple because it promises to our body enjoyment of a physical food.) Even the founder of contemporary aesthetics, Kant had no room for music in his theory of the mind: "(As for) the expansion of the faculties which must concur in the judgment for cognition, music will have the lowest place among (the beautiful arts)... because it merely plays with senses." Kant had no room for music in his theory of the mind, because the knowledge instinct was not known. Hence, differentiation and synthesis were not known. Even today, these ideas remain little known among musicologists; the idea of expression continues to provoke disputes, "embarrassment and confusion."

To summarize, the period of Rococo and classicism in the 18th century turned out to be a cultural respite between the highest tensions of the preceding Baroque and the following Romanticism. In this period a firm synthesis was achieved through a symbol of rational scientific thinking. Let me emphasize that synthesis was achieved not through science, but through the symbol of science, which assimilated projections of the mystery of human soul. Correspondingly, creative forces were directed at differentiation: In politics—at the idea of Enlightenment. In philosophy—at the differentiation of the idea of the mind (Kant). In music—at the differentiation of emotions (Bach sons, Gluck, Calzabigi, Haydn, Mozart, Beethoven).

SPLIT SOUL: ROMANTICISM

Refined Classicism near the beginning of the 19th century was replaced by the unbound Romanticism, why? Changes in consciousness and culture, as previously, were defined by a complex play between the two factors, differentiation and synthesis. Synthesis of Rococo and Classicism based on the rational was destroyed by differentiation of rationality. Once more, as it happened in history many times previously, the very foundation of synthesis has been differentiated gradually leading to its destruction. So, by the end of the 18th century the rational uncovered a complex, diverse, and irrational nature of emotions. Mechanisms of the mind found by Kant, understanding and emotions, for the first time rationally explained mysteries of unconscious, beautiful, and sublime. Kant outlined the directions toward scientific unification of rational and irrational, but his works were too complex for intuitive understanding. Kant delineated future unification of physics and philosophy; however, mathematical methods for describing Kantian intuitions did not yet exist, and much of it was lost for his followers. Kant was too much ahead of his time. Penetration of art into unconscious, the critical gaze into oneself—the method of Kantian aesthetics, only a century later, would become the foundation of Modernism.

The immediate followers of Kant inherited another aspect of Kantian theory, the awareness of *the a priori structures of the mind*. For thousands of years, philosophers were trying to explain contradictions in thinking by arguments. Kant revolutionized thinking by explaining that often mind is not directed by arguments, but arguments are directed by a priori (inborn) structures of the mind. The a priori structures make creativity possible and also set its limits. As a reaction to this self-contradictory nature of the a priori, there is a romantic dream of a pristine, perfect cognition of a primitive human, unspoiled by education and concepts. Whereas Kant discovered a conscious rational approach to analyzing unconscious and its limits, *the romantic idea of creativity was a denial of limits, 'leap through' unconscious*.

In this denial of apriority, romanticism replaced God with the *subjective human ego*. Ego, the subjective, when mistaken for the entire psyche, attracts unconscious projections of the highest concepts of meaning. This explains Romantic contradictions. The romanticism near the end of the 18th century emerged as a political force in the French revolution. Thousands of people died at the hands of Marat, Robespierre, and Danton, romanticists who placed a part of their ego, *conscious ideas* of liberty and brotherhood above all else, above reality of the unconscious and human experience. Jung analyzed the psychological dangers of *inflation of the ego* (Jung, 1948) or "puffed-upness," when self is assimilated to the ego. The unconscious is psychologized, its power is unrecognized, leading to enantiodromia: conscious and unconscious are fighting instead of supporting each other, this leads to a psychic catastrophe. Humankind continues on this dangerous path while a deeper basis for synthesis in the soul is not found. Enantiodromia of the Romanticism led to the WWI, WWII, and in significant extent to contemporary 21st century catastrophes. Romantic ideas of the highest purpose exert profound effects on artistic and political thinking till today. The great achievements and failures of mankind in the 19th, 20th, and 21st centuries have been defined by romantic consciousness.

At the beginning of the Romanticism, romantic creative "break-through" was a source of inspiration and basis for synthesis. Romanticism continued the creative process of differentiation and synthesis, accepting *as the highest purpose and value the subjective, man* with his rational and irrational. Differentiation of notions of the highest turned to human emotions and feelings. Rejection of "material," immersion into "pure spiritual creativity" were often associated with romantic poets and composers. The creative "break-through" was a romantic source of inspiration and a foundation for synthesis for a larger part of the 19th century.

Musical romanticism began, possibly with Hoffman and Shubert. Beethoven connected it with the classical tradition: "Beethoven's music sets in motion the lever of fear, of awe, of horror, of suffering, and awakens just that infinite longing which is the essence of romanticism... (He) opens to us the realm of the gigantic and unfathomable. Glowing rays of light shoot through the dark night of this realm, and we see gigantic shadows swaying back and forth, encircling us closer and closer, destroying us, but not the pain of infinite longing in which every delight, rising up in joyful voices, sinks and drowns, and only in this pain, consuming love, hope, joy, but not destroying it and aiming at bursting our chests with its unison of all passions, do we live on and are we rapturous seers of the realm of spirits!... In the artistic construction (Symphony in c-Minor, the 5th) the most wonderful images pass by in restless flight, in which joy and pain, melancholy and bliss emerge next to each other and intertwined. Wondrous creatures begin an airy dance while flowing towards the light, moving away from each other, glowing here and glittering there and chasing each other in manifold groups; and in the midst of this unlocked realm of the spirits, the delighted soul is listening to the unknown language and understanding the most secret longings which have taken hold of it". E.T.A. Hoffman (1813).

Romantic elements had been present in the collective consciousness for a long time. Even in medieval knighthood, ideals of devotion to one's lord, friendship, and beloved woman, romantic feelings were bestowed with the highest meaning, replacing God. Romanticism in music can be heard in Monteverdi, chromatic organ works of Bach, in his sonatas for clavier and violin, in Handel's expressive arias. "Feeling" became consciously valued when Rococo was at its height, especially in the works of

Bach sons. If looking back at the past periods, one pays attention only at musical style, it is easy to overlook that a fundamental change in consciousness occurred. One may read in a textbook simple facts, like: "In classical music form and order came first, in Romantic music expressive content." But, it is worth remembering that "form and order" or "expressive content" came *first in music because they came first in consciousness*, even if being unconscious at first, music brought romantic ideas in the consciousness.

Let me repeat, romantic synthesis was founded on subjective, bowed to it. Therefore, romanticism emphasized fleetness of human states, their uniqueness, and differences. Musical Romanticism was closely related to nationalism, impressionism, expressionism, and "realism." Let us decipher these terms. Nationalism saw folklore as more valuable than generally-humane; expressionalism emphasized emotions and mental states; impressionalism valued ephemerality as more important than transcendent and eternal. "Realism" had nothing to do and was even opposite to philosophical realism of Plato, Aristotle, Kant, and Nietzsche. According to philosophical realism, the real "things" were concepts in our mind based on mental representations, concept-models, whereas romantic "realism" concentrated on some arbitrary, often lowlife aspects of human experience, as if "more real" than lofty ideals. But this is wrong, reality is constructed in our minds in interaction between concept-models and the world, and "really real" are those concepts and experiences that help us to survive, not as mere pieces of flesh or animals, but as humans, and to achieve maximum realization of ourselves, of our infinite potential, and of course, the highest manifestations in every art form transcend limits of collective consciousness.

The emotional range of music was enriched in the romantic period; emotions became more urgent and intense. Musical forms expressing diverse extreme emotions got freer; the range and number of instruments increased. Orchestra were more widely combined with solo; music became more chromatic, tone color richer, harmonies widened, melodic phrase structure freer, while melody ascended over harmony. Richness of tone color and chromaticism means more systematic deviations from consonant sounds that our ear "automatically" perceives as harmonious and plain. Harmony emphasizes "vertical" space of music sounds, when several tones sound at the same time. Melody is a "horizontal" structure, a sequence of tones. Melody is the most expressive, emotional element in music, possibly because the original melodies were those of human voice.

Characteristic new genres were solo songs with piano accompaniment (Schubert, Schumann, Brahms, Wolf) and symphonic poems, as if musical stories combining emotional with conceptual. This combination was the quintessential Romantic way to achieve *synthesis* between consciousness with its richer differentiated emotions and the unconscious-highest that was identified with the subjective human experience. In parallel, romanticism *differentiated* emotions, created new conscious experiences by bringing unconscious psyche closer to consciousness. These were achieved by the boldness in modulating to ever more distant keys, coloring without a resolution in a perfect cadence, which resulted in "chromatic frustration" for the listener.

Violent burst of romanticism replaced balanced harmony of classicism—could a scientific theory of consciousness follow these seemingly unpredictable movements of collective soul? Let me remind that an ideal harmony is just a first step toward beauty. An ideal harmony corresponds to rational consciousness of classicism. However, consonant sounds, like psychological equivalents of perfect correspondence between concept-models and surrounding world, cannot satisfy the KI or refined musical taste. Classical harmony self-destroys in differentiation of emotions. In consciousness, rationality of classicism is superseded by irrationality of romanticism. In music, passions are coming to the fore. Passions in music are created by contrasts between consonances and dissonances. The more composers wound up the dissonant tension, the clearer one felt a psychological need for its resolution in a consonance.

"By the 19th century, composers knew and frequently employed elaborate strategies for delaying resolution and, concomitantly, increasing tension. This, along with inflection, chromaticism, and modulations makes music sound 'expressive'".

Let me remind, tonality was founded on alternating tensions and resolutions; this required chords to be defined on certain scales determining the direction of resolution and psychologically perceived as the expected "direction of motion" of musical sounds. Inflections and chromaticism are deviations from the psychologically expected direct progression of tones; modulations are changes in the background "scale" of the tone perception. The tonal system was created over hundreds of years for expressing emotions, and the single most important tonal sound in music was triads. Triad is a three note group of tones, like do-mi-sole, made up of a major and a minor thirds, diminished and augmented, their inversions, etc.

"In the Romantic period, the triadic system was exploited to the farthest consequences, chromatic modulations and distant keys... diluted the strength of a single tonal center and tonality started to disintegrate." Although the basis of tonality was in the emotionally distinct desire for returning to the tonic, to resolution, more and more complex *unresolved emotions* were gaining importance during the romantic period, emotions in chords without a clear direction of resolution.

The "diminished seventh" chord, in the Beatles "Michelle" is the fourth chord of the song, "Michelle, ma belle." This chord lacks a tonal center and sounds ambiguously (as a major and minor at once without a clear "direction of resolution" of its tension). Diminished seventh was discovered by composers long ago. Psychologically it might be perceived as a "climax." Bach used it this way. During romantic period, when inspiration was thought in irresolvable states and emotions, Liszt and Chopin amplified its psychological tension making it sounds as "impending catastrophe." It is among the main "personages" in Tchaikovsky's Sixth symphony; Mozart, Weber, Verdi, and Wagner used it to evoke a feeling of terror or suspense. Beethoven's 32nd Sonata in C minor Opus 111, wrote Jane Coop (2016), "Opens with an intensity of passion which transcends anything Beethoven composed previously, the first lightning-bolt chord... a diminished seventh... sets the character for the whole movement—one of torment and strife, carried along by hard, driving energy."

To emphasize connections of consciousness and musical forms, let me repeat that during romanticism the highest values were spontaneous emotions and experiences, serving as sources of inspiration. New psychic states and extreme emotions were created and expressed by ambiguous chords with uncertain directions of resolution, transcending limits of the tonal system. The famous Tristan chord in Wagner's "Tristan und Isolde" is even more ambiguous than the diminished seventh. Saturation with dissonance and tonal-ambiguous chords without resolutions in closed cadences created the Wagnerian unending melody filled with the finest colors of irresolvable emotions.

Artists and composers always attempt to influence the ways of history, the objective, the super-personal, and this is why unattainability of human aims became the main object of romantic art. An ideal of the subjective, which the Romanticism placed on the highest pedestal, came into contradiction with itself—because the meaning of human existence transcends the limits of *Ego*. This is why not repose, but restless seeking and impulsive reaction governed romanticism; it sought out a radical kind of subjective expression, the new, the adventurous, concentrated on remote and strange, was haunted by a spirit of longing. There were boundlessness, unbounded emotions, movement, passion, and endless pursuit of the unattainable.

One cannot, however, "stop at Self." Bowing to the subjective, romanticism laid the foundation for its own demise—differentiation of the subjective in the finest works of art revealed the internal limitation of romanticism—the subjective disappears and dies inside itself. Romanticism valued an individual human being above all, but it threatened the individuality in a psychic process that Jung called psychological overvaluation, "puffed-upness," which led personal lives of many romantic poets and composers to tragic demise. The "end of tonality" was not due to its exhausted means, but because subjective romanticism destroyed itself.

To reiterate a premise of this chapter: *differentiation destroys synthesis that inspired the differentiation, and this contradiction is the basis of the cultural evolution. The highest romantic achievements in music—differentiation of emotions were destroying the tonal system created for expression of emotions.*

CONSCIOUSNESS AND MUSIC IN THE 20TH CENTURY

Changes in consciousness near the end of the 19th century affected all directions of cultural life, including politics, science, philosophy, arts, and music. Ideas inspiring a creative person changed. In politics, an idea of a national state was superseded by "brotherhood of humanity" as the highest goal; in economics, individualistic capitalism was replaced by socialism and communism; scientific inspiration—by formal logic; Hegelism—by positivism; poetic symbolism—by "realism"; and impressionism in paintings—by abstractionism. The romantic idea of a free *individual*, unbounded, creative was replaced by an idea of society. Subjective intuitions and emotions were displaced by objectivity, by creativity free from personal emotions and intuitions; challenging society—by egalitarian equality; beautiful—by disgusting; and "unique"—by the "mass-produced." Collective consciousness idolized an idea of the *objective*.

In music this tendency *away* from individual and *toward* the objective was most clearly formulated by Arnold Schoenberg. He thought that the traditional tonal system developed over centuries for expressing *human* experiences, emotions, and intuitions, was "no longer available, as a vehicle for sincere artistic intentions... (diminished seventh) fell from the higher sphere of art music to the lower sphere of music for entertainment... a sentimental expression of sentimental concerns. (Diminished seventh) became banal and effeminate." I remind that the diminished seventh chord was used frequently by most great composers over the previous two centuries, it sounded as a "climax" or "impending catastrophe."

To overcome the psychology of tonal music, Schoenberg formulated an atonal rule, or 12-tone technique (dodecaphony): A composer had to use all 12 notes of the scale (all white and black piano keys), before any key was repeated. This arbitrary rule (with few modifications) played a major role in 20th century music. Why? Whereas in social life *synthesis was coming from the objective* (socialism, communism), there was no such idea explicitly formulated in music. Still many composers were literally "lost" among the unlimited possibilities of combining sounds. Schoenberg's rule limited unboundedness of possibilities and gave direction. This *limitation of uncertainty* some musicologists viewed as the main reason for attractiveness of Schoenberg's technique for composers, it seems there is some relation of the equality of tones to the idea of equality of people creating a synthesis in social life.

Dodecaphonic or serial direction in music was similar to contemporary directions in linguistics, philosophy, and mathematics. Linguistics was dominated by Saussure's idea that the meanings of words in language are defined through other words (like in a dictionary), and that relations of words to objects in the surrounding world are arbitrary and defined by conventions. Saussure rejected Humboldt's subjective creative "inner from" of a word—what today we call synthesis of conscious and unconscious. Philosophy was ruled by logical positivism, maintaining that all metaphysics, the entire totality of knowledge exists in a form of relationships among notions, that *content is form*, and no other meaning exists but form. In psychology, behaviorism emerged explaining all human behavior as a sequence of stimuli and reflexes, and denying consciousness as a necessary scientific notion of human intellect.

All these developments were inspired by and related to the idea of the *objective* and its manifestation in mathematics. In mathematics and logic, the objective was manifested in formalism: The only meaning of the mathematical objects was defined by their logical relations and arbitrary axioms. The quintessence of the formal direction was formulated by a mathematical logician David Hilbert, in his famous attempt to formalize mathematics and logic (1900). Hilbert and his followers were searching for the laws of human thinking—they wanted to formalize the entire human creativity, as if the human mind and creativity consisted of predetermined rules. The same type of consciousness was reflected in Schoenberg's ideas.

The formal direction seduced many by its novelty, and still it is "as old as the world"—it is nominalism, a type of consciousness that appeared two and a half thousand years ago in an old dispute in classical Greek philosophy between realism and nominalism. Realists thought that the principal reality were ideas, concepts of the mind, they were inborn and learning was secondary. Nominalists denied reality of ideas, equated them with words, and thought them to be arbitrary signs as in formal logic. The entire knowledge they thought is learned from experience. Nominalism expressed in terms of the 20th century became logical positivism and formalism. Divide between Realism and Nominalism was overcome by Aristotle. In his theory of the mind, inborn ideas existed as vague potentialities; they turned into logical crisp actualities "in interaction with the world." But these fundamental aspects of his thoughts have been lost through the centuries and only today we are becoming aware of them. Aristotle thoughts have been amazingly close to dynamic logic of the contemporary science. Aristotle as well as contemporary science resolved this issue: both inborn *"a priori" structures of the mind and learning from experience* are necessary. Science requires theoretical predictions and experimental confirmations. The KI combined realism and nominalism. Notwithstanding, philosophical disputes continue under different names. Why?—Jung uncovered *psychological* basis for continued disagreements: The two opposing views of the world are related to the two opposing types of psyche, introverted people tend to realism and extroverted tend to nominalism. Scientific exploration requires combining realism and nominalism within the scientific method.

But let us return to evolution of consciousness and music near the end of the 19th century and the beginning of the 20th century. Formalization in this period became the source of synthesis and main method of search for the objective. In mathematics and logic the precision of proofs was enhanced; music concentrated on seeking new forms. Playing with forms as a self-sufficient purpose, however, distracted creative interests from the deep processes in human psyche and from mechanisms of the mind, whereas "only content can be innovative."

Gradually, infatuation with the objective self-destructs; synthesis inspires differentiation, which destructs synthesis. In the 20th century this cycle tremendously accelerated. In mathematics, Gödel used perfected precise tools of logic to prove inner contradictions of logic. Logic turned out not to be omnipotent, not as "logical" as expected; logic contradicted itself and could not explain the mind. Soon thereafter (in the 1940s), logical positivism in philosophy disintegrated. In linguistics—mechanisms of the mind and *not the forms* of language determined new directions. In psychology—the formal direction of behaviorism lost attractiveness since the 1960s. In music—serialism and other directions closely related to the idea of equivalence of form and content also lost attractiveness (although, it still attracts interest among some academic musicians, despite public indifference).

A contradiction between form and content comes to light in the creativity of the very founder of musical formalization, Schoenberg. Although he suggested formal rules of dodecaphony aimed at bringing the human spirit beyond limits of emotions, his own works strived to create sublime human feelings. He wanted to express in music what is inexpressible in words, the Biblical prohibition of creating God's images. Unverbalizable nature of the Divine he wanted to express in music. As known, Schoenberg worked on this theme for more than 16 years, on the oratory "Jacob's Ladder," then on the opera "Moses and Aron." Schoenberg, however, came to a conclusion that he cannot create music adequate for this high aim. Both works remained unfinished. The inner logic of his failure is possibly in the principal contradiction between Schoenberg's goal (to create in music a purposeful meaning) and arbitrariness of dodecaphony, its detachment from the world of human experiences (as formulated by the composer himself).

Could dodecaphonic, serial, atonal music express human experiences, and what kind? Possibly the first convincing example of atonal music expressing human passions belongs to Schoenberg's pupil, Alban Berg. His opera "Wozzeck" is considered a high point of 20th century music. But I would like to ponder, if Berg achieved what eluded his teacher. Did Berg use atonal music to create new meanings, new feelings inexistent previously? What are these new meanings? The opera is about a man who lost belief in himself, in love, and commits suicide. Atonal music in the opera expressed extreme hysterical harrowing human states, inexpressible in any "beautiful" tonal music. "Wozzeck" could be compared to contemporary thriller movies, which went much further in this direction, and use music that often cannot even be written with notes. Hysterics, of course, was not a psychological discovery, but its expression in music was discovered by Berg.

Kierkegaard, Dostoevsky, and Nietzsche brought into collective consciousness an idea of unsolvable existential tragedy of every person contained in the finiteness of life. Whereas in an Ancient Greek tragedy spectators were attracted by mystical inscrutability, whereas listening to Bach I feel emerging new previously unknown sublime feelings, giving meaning to my life, what attracts in Wozzeck? A spectator sympathizes with Wozzeck, condescends to a suffering little man. It reminds one of "curiosity factors" on a highway: Traffic slows down near a car crash, spectators want to see the crash. People want to sympathize with victims. Compassion raises man above animal. But after Palestrina, Shakespeare, Bach, Gogol, Nietzsche—Wozzeck was not a discovery in human consciousness. Alienated compassion continues to affect every sensitive person; however, this way of feeling was *brought into consciousness* long ago by 4000 years of monotheistic religions.

And there is another view on "Wozzeck." Although Bach inspires human to reach to God, Berg perceives the meaning of life as unavoidable defeat in a fight with the devil. The 20th century, possibly more than any other period, was a colorful illustration of this fight. Two world wars, millions of prisoners of fascism and communism, hostages of terrorism, indeed a cannibal backdrop of the century. This backdrop is a part of contemporary consciousness. Of course, evolution of music and art corresponded to this consciousness, and we see it in "fighting" differentiation and synthesis. Long-known harrowing states of psyche have to be comprehended within contemporary differentiated consciousness.

While the dodecaphonic formalization emerged in music, a similar objective direction in poetry was developed by T.S. Eliot. He was freeing poetry from romantic ways; he searched for this freedom in impersonal poetry, *depersonalization*: "The progress of an artist is a continual self-sacrifice, a continual extinction of personality." In depersonalization Eliot saw art coming close to the objectivity of science. Similar doctrine in art, called dehumanization, was studied by y Gasset. Dehumanization of art deleted the human figure, insisting that form and not a metaphor of human defines contents of an artwork (I will repeat, these directions in music, poetry, and visual arts followed logical positivism, emphasizing identity of form and content). "Art purged of those 'human, all too human' elements that to artists in the early 20th century suggested ephemerality, inconstancy, mortality, in favor of abstract patterns and precision suggesting transcendence of our muddy vesture of decay..." In creative arts this purge meant a move away from content and expression. A similar distinction was made between vitalist art—which imitates the "forms and movements found in nature"—and objective geometric art, which relied on "something fixed and necessary." Igor Stravinsky, 1939–40, similarly insisted on "the higher mathematics of music."

What are the meanings indicated by objectification, equalization of form and content, *rejection of human*? A poet, Nobel laureate, and American National Poet Laureate Joseph Brodsky as if responds to Elliot. "To start, only content could be innovative... Poetry is not an act of self-elimination... poetry is an imperative art, thrusting its reality in the reader. A poet striving to demonstrate his ability for self-elimination... shouldn't limit himself to neutrality of diction... he has to make the next logical step and shut up completely... Illogicalism, deformation, pointlessness, disharmony, incoherence, arbitrariness of associations, flow of unconscious—all these elements of contemporary esthetics... theoretically are called to express its peculiarity—in real life are categories of marketplace, which is witnessed by the corresponding price list... Atomization and brokenness of contemporary consciousness, tragic attitude, etc. seem today so conventional a truth, that their expression and usage of means demonstrating their presence, turned... simply into the demand of market. The demand of innovativeness in art witnesses... the dependence of art on the marketplace reality."

Alongside with objectification, Modernism includes its opposite, critical gaze into self, self-reflection. It breaks into unconscious, subjective, and objectifies the inner world. Penetration onto the depths of unconscious, however, led to the subjective and contradicted to an idea of the objective, the basis of synthesis in Modernism. *Again, the fight between differentiation and synthesis was reenacted. Differentiation of an idea of the objective in Modernism destroyed synthesis. By opening depths of the unconscious, the best modern artworks revealed complex interactions of the objective and subjective, and proved that creativity cannot exist limited by the objective.*

Modernism in collective consciousness begins with Kant, who initiated scientific psychology and even cognitive science in a prescientific era. The main aspect of Modernism is self-reflection, critical gaze into self, the *objective* gaze. However, penetrating into unconscious, the best works of modernist art revealed complex interlacing of the objective and subjective—creativity cannot exist within bounds of the objective. Investigations of unconscious depths led to subjectivism and *contradicted* the idea of the objective, which was the basis for the 20th century synthesis and inspiration. Differentiation of the objective in Modernism destroyed this synthesis. Postmodern returned to the objective, even at a price of simplifying conceptual content, to message, or no content at all (e.g., music of J. Cage). A postmodern artist concentrated on "emptiness" of any object, on its accessibility to consciousness. Although Modernism continued differentiation at the expense of synthesis, postmodern rejected differentiation to save synthesis, even if simplified, even if empty of any content. The divide between conscious and unconscious in collective psyche, between differentiation and synthesis in modern and postmodern, extended over the entire 20th century.

The very idea of *objective art* contained antinomy manifested in the most unexpected ways. Malevich declared the aim of Suprematism—to free art from any symbolic content—but his "Black square" was interpreted as a symbol of impenetrable unconscious content. Schoenberg formulated the objectified atonal technique—his music however, as the music of his student Berg, as already mentioned, was directed at subjective contents. In "Ulysses" Joyce created a form of language to express a "stream of consciousness," but an almost complete absence of consciousness was the outcome.

Jung characterized Joyce's "Ulysses" and along with it a significant part of 20th century art and collective consciousness in the following way: "...A passive, merely perceiving consciousness, a mere eye, ear, nose, and mouth, a sensory nerve exposed without choice or check to... a stream of physical happenings... The stream... not only begins and ends in nothingness, it consists in nothingness. It is all infernally nugatory... Today it still bores me as it did then (in 1922). Why do I write about it?... Joyce has exerted a very considerable influence on his contemporaries... Destructiveness seems to have become an end in itself... (it) is a collective manifestation of our time... *the collective unconscious of the modern culture*... the modern artist immerses into destructive processes, to affirm in destructiveness the unity of his artistic personality... We still belong to the Middle Ages... For that alone would explain... why there should be books or works of art... (like) 'Ulysses'. They are drastic purgatives... for the soul... which is of use only where the hardest and toughest material must be dealt with."

I would emphasize that this analysis is directed not against art contemporary to Jung's writing, but at psychological mechanisms of its perception. Art of the 20th century is such as is its consciousness.

There are no reasons to forget fascism, communism, terrorism... Individual consciousness is still a rare destiny of few people in some moments of their lives. Let us hope that Jung along with art masters would help us to get out of The Middle Ages and free our souls from "the hardest and toughest material." Those agreeing with Jung would find many examples similar to Joyce in music and other arts.

In music, like in literature, visual art, and philosophy, two contrary historical tendencies of evolution of consciousness collided, differentiation, and synthesis. It is not surprising that changes in musical forms paralleled other arts, philosophy, and science. Consciousness, accelerated by scientific method, became self-referential since Kant—i.e., it turned to the analysis of self, to differentiation of unconscious. "A gaze toward self" indicated the essence of Modernism. Differentiation of self, as a penetration into the depths of unconscious was manifested in the psychology of Freud, literature of Joyce, paintings of Pollock, music of Scriabin and Shostakovich (to name just few). However, differentiation destroyed the wholeness of the world perception, and a contrary tendency emerged, postmodern, as a striving for synthesis based on the simplest notions.

Although in the past centuries differentiation may have dominated one epoch and synthesis another, in the 20th century all mixed up. While Modernism sought depths of self, Postmodern with equal force rushed to simplicity of the bases of aesthetics. The opposing tendencies of collective consciousness were present in conscious and unconscious of an individual composer or artist, sometimes as unifying and inspiring, sometimes as pulling apart and destructive. This process, which Jung called enantiodromia, for a moment annihilated the art object.

A distinction was lost between art and nonart. An important role here was played by politics and ideology. Art did not lose value in communist USSR, nor in fascist Germany. Fascists and communists used the power of music for propaganda, similar to ISIS today. As a reaction against this exploitation of art, academic art in the free world repudiated emotional and affective music. Therefore, unexpectedly, in the free world politics strongly influenced the direction of art: "Ideology of cold war... sanctioned association of logical positivism with democracy and formalism with defense of political freedom," wrote Richard Taruskin. The split between conscious and unconscious is painful, because, I repeat, wholeness is an instinctual need, a demand of the KI. To achieve wholeness, men and women are ready to reject diverse knowledge and to narrow their consciousness to frames of ideology—communism and fascism were such attempts to achieve ideological synthesis. Communism promised to satisfy instinctive unconscious needs through the ideal of the objectively sublime (ideal society, Heavenly kingdom on Earth), and fascism promised to satisfy ideals of the sublime through objectivity of collective unconscious instincts. Both attempts led to national enantiodromias: in fascism as in communism, collective consciousness and collective unconscious rushed in opposite directions, rousing along the way world catastrophes and destroying national psyche. As a national phenomenon, the bridge between conscious and unconscious built by fascism survived a little more than 10 years, the communistic "bridge" survived for about 70. However, as local phenomena, there have always existed attempts to reach synthesis using narrow-ideological symbols, cutting consciousness down to simple formulae. Examples include European and Russian terrorists, destroying romanticism of social struggles in the 19th century, and contemporary fanatical and terrorist groups. Today Russia and China having learned from their socialist past attempt to build more human-like societies. The Western world moves in the opposite direction, despite catastrophic experience of socialism in Germany, USSR, China, and other socialist countries, the Western world moves toward socialism. An idea of socialism may seem much more attractive than capitalism, especially for young people: free health care, free education, the ideas of equality, togetherness may create synthesis. It might be difficult to imagine what would be the reality of socialism after wealth accumulated in the past would be gone. And one may think everything would be fine.

In totalitarian societies, music is used to wire feelings. But in pluralistic societies, destinies of art do not seem less troubled. Many people do not want to know unpleasant truths. Existential spiritual sufferings of individuation of a separate human being are muffled, cannot be heard. Concepts of consciousness are losing connections with unconscious instinctive bases of psyche. Mass culture may prefer not to know.

In 2002 British songwriter Mike Batt released an album containing a track called A One Minute Silence. This provoked a lawsuit. "Silence" have already been discovered and copyrighted. The suitor was the estate of a composer John Cage. The case was settled out of court for a large undisclosed sum. This story is not just ridiculous or crazy but also a result of strange and twisted evolution of understanding, or misunderstanding of the relationship between art form and content during the 20th century. Can two art pieces of a similar form (silence) have different contents?

Existential content of a musical piece came to oppose the form, as if form without content had any meaning. Discussion about form and content excited many a mind. Whereas romantic art "acknowledged clear separation between content and form," positivism considered content to be "a function of form." The principal difference we see in that formal rules and structures do not correspond to the surrounding world or the inner world. Sometimes it seems the only tie of Postmodern art to the inner world of man is assumed in the very form itself, as a logical game, as Lego, as a "Game of Glass Beads." Especially destructive is formal innovation in music, because semantics is contained first of all in the sound. "The very phenomenon of rhyme points to existence of interconnections of concepts and phenomena... sound... is a form of cognition, a form of synthesis."

While concentrating on formal innovations, music was losing existential contents. Cage in the 1940s composed music by casting lots; eliminating a creative composer's choice, he came up with several ways of random "divination" in music. Drawing lots as a way of finding the highest purpose is not new. It is a premonotheistic type of consciousness. It appeared when ancient fused synthesis broke down, while monotheistic synthesis did not exist yet. Divination was popular between 1300 and 650 BCE in Mesopotamia, and in some polytheistic cultures it is still used today. At those times differentiation began, fused primordial synthesis broke down, and humans were lost among a multitude of thoughts. Similarly, in the 20th century, scientific thinking accelerated differentiation, synthesis was lost, and composers were lost among a manifold of sound combinations. So, Cage's divinations were a return to consciousness outdated by 3000 years. Later, Cage came up with even more primitive synthesis, prehuman undifferentiated ability *to hear* just anything. "Four minutes, thirty three seconds of silence" reminded "Black Square" by Malevich. But, whereas Malevich as if intended a challenge, an existential gaze into human depths, Cage and his listeners ravished in their ability to perceive at least silence. Cage's primitivization of musical consciousness continued a process started by Berg—expression of primitive emotions in modern musical form, emotions much more primitive than psychology of a catastrophe in diminished septachord, called "banal" by Schoenberg. Arbitrariness of the atonal Schoenberg rule was psychologically close to random divinations by Cage; both methods witnessed a loss of meaning.

Further elimination of meaning in formalization of music was developed by Pier Boulez. The rule of serialism, nonrepeating notes until the series was used, Boulez applied not only to pitch but also to other elements of musical organization, like rhythm. Boulez was interested exclusively in the form of a work of music—"the game of glass beads." He criticized Schoenberg and Berg for "inconsistent" serialism, for their desire to use the new form for expressing "outdated" content—human emotions. Boulez represented extreme formalization in music, he denied the human, the emotional, and he was called the "murderer" of melody. Being more active as a conductor than composer, Boulez excluded from repertoires of orchestra under his control any music reminiscent of melody. As chief conductor of the BBC Symphony Orchestra and the New York Philharmonic, he refused to perform Mozart, Tchaikovsky, Prokofiev, Shostakovich, Britten, Poulenc, and Schnittke.

I named just few musicians attaining the heights of recognition in the Western world in the 20th century. Does it mean that the culture has already disintegrated? No, not yet, but it might.

Poststructuralism was a direction in postmodernism, considering the entire culture as consisting of "empty codes," which only referred to each other and denied any true "metaphysical" reality. Poststructuralism and closely related deconstruction equated content with form and denied any other meaning but form. Poststructuralism in music (like music of Boulez) meant that tones and other elements of musical structure acquire meanings only in their interrelationships, and not in their links to the world of emotions, thoughts, and inner human life. Music was prevented from entering the gates of its main purpose—restoration of synthesis of conscious and unconscious. Poststructuralism was immersed into a complexity of play, simplified in human content, and to such an extent rejected the rest of human culture that its disappearance cannot even be called a crisis, it just disappeared.

Let us return to the analogy between mathematics and music. Pythagoras saw a *number* as a mystical object, a source of synthesis, direct manifestation of the divine forces acting in the world. Therefore, connection of musical harmonies with numbers Pythagoras perceived as pointing to the divine *power of music in the world*. During the past two and a half thousand years consciousness differentiated, and for a mathematician today numbers are simple axiomatically defined objects. A mathematician sees mystery in the very *possibility* of a theory, of a theoretical correspondence-harmony among complicated mathematical objects that cannot even be described in words of language. Some musicians, on the contrary, have been attracted by the mystery of numbers, its *objectivity*—as a substitution for the human world of majestic and tragic, too complicated as compared to the simplicity of numbers. "My music," Stravinsky wrote, "is far closer to mathematics than to literature... to something like mathematical thinking and mathematical relationships. I don't assert at all that composers think by equations and tables, or that such things can symbolize music. But a composer's way of thinking—the way I am thinking, it seems to me is not very different from mathematical." "Mathematical thinking" that captivated composers and musicians in the second half of the 20th century reminds of idolatry, and Stravinsky's confession in full measure refers to Boulez, who was inspired by tables of numbers. A mathematician however thinks by concepts, not by tables or equations (it might be that an F-student, when failing an exam, thinks by numbers and tables). *Best mathematicians are attracted to mathematics by mystery, whereas composers—by its absence.*

Apparently, musicians were looking for objectivity in mathematics, and "pure" play of forms reminded them of mathematics. Some so-called "serious" academic musicians, looking for the *objective*, rejected human content, repudiated serious emotional content. However, synthesis based on the objective was broken down by differentiation of the objective, and the very possibility of a serious work of art was often eliminated. As a consequence, another form of postmodern, minimalism searches for minimally representative forms, primitively imitative (noise, silence), or "mathematical" tables of tones, rhythms, articulations—complex for musicians, but meaninglessly simple for mathematicians. "Serious" music ceased affecting anybody, except for narrow specialists interested in formal play, although intellectual tension of such games could be fairly strong, like in chess.

Search for objectification in music proceeded in two directions, first—serialism, antiromantic romanticization of formal. And second—romanticization of breaking out beyond the limits of the subjective human; in the opinion of Taruskin it was developed by composers of the Russian school. Stravinsky was inspired by romanticism of the primitive; in "The Rite of Spring" one hears synthesis in prehuman primitive, which objectivity is in biological foundations of prehuman passions. Scriabin searched for inspiration in an idea of the highest objectivity, exceeding all human desires; his synthesis sounds in a mystical chord of pleroma in "Prometheus." This chord is more uncertain than Tristan chord, more ambivalent toward any direction of resolution.

"The pleroma, a Christian Gnostic term… for the all-encompassing hierarchy of the divine realm, located entirely outside the physical universe, at immeasurable distance from man's terrestrial abode, totally alien and essentially 'other' to the world of concepts and whatever belongs to it".

Shostakovich has shown the antihuman nature of "pre-" and "post-" human, and revealed the impossibility of creating synthesis outside of human nature. Shostakovich's tradition was continued by Schnittke, making music speak in the language of existential tragedy. This is why Schnittke's music is so full with tints and shades of irresolvable sadness.

Today this tradition of tragedy and sublime is continued by Sergei Slonimsky. It might be because Russian language in its grammatic structure keeps a testament of connected emotions and concepts, synthesis which Russian culture has contributed to world culture.

Looking back into the 20th century, future musicologists will less value formal innovations of Schoenberg and other atonal and serial composers, who "emancipated harmonic dissonance," but will recognize as a most important contribution music of Shostakovich, Prokofiev, Schnittke, Stravinsky, Britten, Hindemith, Poulenc, Orff, Miyo, Honneger, Messiaen, Bartok, Janacek, Penderecky (to name just few)... who "emancipated semantic dissonance." This Taruskin's prediction comes amazingly close to the meaning of music in overcoming cognitive dissonances as discussed in this book. Aesthetics of the way "art communicates with its audiences, will again replace author-centered poetics as the primary object of study." Musical profession will come to value originality "*not in the way that their music sounds, but in the way that it means.*"

A division between popular and serious in 20th century art and especially in music was amplified, as if objectively existing. An ever-increasing role in contemporary musical culture belongs to popular music. Is there an objective difference between popular and serious? Evolution of consciousness driven by complex interaction between differentiation and synthesis does not follow a straight line. And if in the Middle Ages new forms of music creating strong emotions were regulated by Church, so in the 20th century this role was taken by some "academicians" rejecting music of human passions. As discussed in this book, human passions are the essence of music. Still, some musical theoreticians are ready to assign emotional music to ideology, or entertainment and popular, and ban it from high art.

Separation of popular from serious always was a difficult task. In the past a default solution often was given by technology and affordability: marketplace art was popular, while churches and castles were places for the serious. Was it really so? This issue became more complex in the 19th century. In the 21st century the problem became even less solvable. Mass culture became a new phenomenon propelled by industrial innovation and commercialization of art. Popular and serious became completely mixed up. Mozart was popular. The best works of Presley, The Beatles, and Weber, combining Mozart with Africa, seem more serious than ideas of Cage outdated by 3000 years. But, how should admirers of serious music interpret the fact that pop culture attracts hundreds of times more attention than Mozart ever had? How can we understand that mass culture belongs to young? Is it even true? And what will be the consequences? Public is buying music since Renaissance, for 500 years public tastes determine directions of music. But this influence changed drastically in the 20th century. Young people spend more money on music than any other sector of the society. Musical mass culture is measured by billions of dollars. Tremendous number of singers and groups vie for the public attention and overfill multiplying numbers of stages, platforms, scenes, CD and DVD shelves, TV channels. Center of mass culture music is the United States. Export of music from the United States exceeds that of cars, tanks, and planes. Not everyone likes this music, and there is a lot of very different music. France adapted laws limiting import of American culture. But "What's coming no padlock can check." Is the aim a preservation of national culture? Or pop-culture industry affects international politics? Industry producing CDs, DVDs, guitars, microphones, amplifiers, players, electronic media, and communications aimed at music need fans and have to produce them. Music is better than tanks, and music, it seems, may have a better chance to overtake the world than tanks. Should we assume that the future of the world culture, the future of history is bright and sunny? Are talents thriving and culture evolves? Can a talented person break into a multibillion dollar industry?

Recently I was watching an interview with an American rock star on a Russian cultural talk show, "Night flight" a Moscow equivalent of Charlie Rose. The show host Andrei Maximov asked the star about what is important to achieve success. In the traditions of Russian culture, Andrei as well as his viewers expected something about talent and hard work. Instead they heard that one needs a few million dollars. "Is not even talent needed?" The star replied that a lot of people have some talent. If you have enough money to promote yourself on TV, day and night, within a month you'd become a star.

Despite this cynical aspect of show business reality and mass culture, outstanding talents overcome artificial divisions between "serious" and popular and make serious music interesting for millions of listeners; we can see this in musicals of Webber and Tanonov. I would challenge artificial divisions between a musical and opera, as one addressing a mass listener and another addressing a serious one. I would continue Taruskin's thought that a cultural role and influence of a music style is defined not by artificial rules, but by the new contents, which is the only aspect of the piece of art that can bring innovation.

Another aspect of mass culture and huge audiences it reaches is a dependence on success that might be impossible to break for a creative person. Success could be a source of inspiration, but too often it becomes a drug, requiring additional stimulations with external opioids, which leads to death. Many rock stars died this way. The role of success in opioid addictions has not been studied. Is it a new development in cultural evolution?

Looking at mass culture within the theme of this book, coevolution of consciousness and music, it is clear that humankind is not yet ready for individual consciousness. In the contemporary life, pulled apart by fast multiplying differentiated concepts, a human being is missing synthesis. This is what music brings to its listeners, synthesis. This is why a young person today spends so much time listening to music. He or she needs to make sense out of the fast changing life, needs to find "the vector of the direction" to the meaning and purpose. Human soul cannot exist without synthesis. What could be the foundation for synthesis today, five centuries after Reformation, three centuries after Newton? What could create a unifying process-symbol in contemporary psyche? It seems that the most survivable, vital inheritance from the 20th century is not Modernism or postmodernism, but the idea of mass culture. This idea unified strivings of every person to realize his or her self here on the Earth, during the lifetime. These strivings could be dangerous. Sometimes, a creative person becomes one with a crowd and for a moment realizes this dream. Feeding itself, this dream may turn into Uroboros, a snake swallowing its own tail, the idol and the crowd.

Mass culture is a logical step in evolution of consciousness, in interaction of differentiation and synthesis. Achieving their unity in individuation is difficult. There is a chasm between differentiated concepts existing in culture and capacity of a single person to assimilate this culture, while preserving synthesis within one's soul. Is this chasm unprecedented and unique for our times? Was this chasm smaller for Aristotle and Ancient Greek crowds? Surprising animalistic and satanistic styles of some rockers and rappers could be understood if we compare them to dithyrambs of satyrs. Remember that the dithyrambic chorus of satyrs was an ancient way of creating synthesis, connecting the sublime with bestial unconscious bases of psyche. The rift between conscious and unconscious threatens the death of culture and "demands restoratory sacrifices." Rap is a contemporary dithyramb, restoring the connection between conscious and unconscious. In both dithyramb and rap—quite regular thoughts are cried out at the edge of frenzy. As in Ancient Greece 2500 years ago, so today in a complex multiform culture, people, especially young people, are losing their bearings. Words no longer call forth emotional reactions, their prime emotional meaning is lost. By shouting words along with primitive melody and rhythm, a human being limits his or her conscious world, but restores synthesis, connection of conscious and unconscious. An internal world comes to wholeness, reunites with a part of the surrounding culture.

As postmodern was a return to pre-Aeschylean, Apollonian consciousness of pure notions—so Rap is a natural continuation of postmodern: Dionysian breaks forth into Apollonian consciousness. These types of consciousness antiquated about 2500 years ago. But consciousness does not whirl in a closed circle. Conceptual and emotional contents of contemporary culture have become much richer, and the previously unseen poles of differentiation are to be unified by the coming synthesis.

CONSCIOUSNESS, MUSIC, AND CULTURE

This chapter concentrated on historical changes of musical forms, which paralleled evolution of cultures and emergence of *individual consciousness*. Philosophers of the past might have called it historical realization of the principle of individuation. To emphasize this main topic, I have set aside a score of problems. I leaned upon scientific understanding of the mind functioning to trace changes of forms of consciousness parallel to changes of musical forms. Summarizing, I would emphasize that music is a most mysterious thing on earth; it contains differentiating and synthesizing powers. Necessity governs relationships between these powers: when rocking toward differentiation, concepts lose meanings, but when rocking toward synthesis, strong emotions nail down thoughts to traditional values. Both lead to a slowdown of cultural evolution. As no other arts, music can forestall cultural slowdown. Music transports reality into the hearts of listeners and restores a possibility of continuation of culture. But will the unity of differentiation and synthesis extend over life? Or will our entire culture be torn into shreds?

From the last prophets and the first philosophers to the age of the Renaissance, a culture was created based on rational thinking and the religious symbol, which did not exist in previous millennia; the symbol of suffering God. *This mystical symbol* gave meaning to human existence by assimilating and expressing unconscious feelings of the tragedy of human existence. Since the Renaissance a symbol of human *reason* displaced the old mystical symbol; roles of individual and rational were strengthened in consciousness. Science accelerated this process, and a new myth of *the rational scientific* mind emerged. Content of the ancient mystical symbol—the existential tragedy—was transferred into consciousness. Mystical inspiration was lost and the Reformation acknowledged this fact. The Church was no longer an intermediary and man faced God alone. *A tension between spirit and matter reached strength, which had never before been experienced in history—and the connection between conscious and unconscious broke down.*

A continuously winding whirl of collective mind processes followed in, which ever accelerated changes of forms of consciousness in search for new inspiring ideas. But whereas the ancient mystical symbol had unified contents of conscious and unconscious, such synthesis cannot be reached now by new rational symbols. Rational foundations of science and mind contradict to mystery of the ancient symbol, which has formed the basis of synthesis.

New symbols bear contradictions close to the surface of consciousness. Correspondingly, new ideas of *the highest* bear causes of self-destruction from the start. For example, classicism of the 18th century replaced God with the *rational.* But differentiation of this idea led to its destruction—emotions turned out to be irrational. Romanticism in the 19th century was inspired by the idea of *subjective humane.* But differentiation of the humane destroyed the romantic symbol—too much inhumane was inside the romantic human. Modern and postmodern worshiped the idea of *objective,* but differentiation of the objective uncovered its limits and destroyed synthesis—a human being cannot exist limited by the objective.

Part of the reason could be in that new symbols still keep separately science and mystery. After the Renaissance, and even more so after Newton, collective consciousness believed that science is rational (nor Newton, nor Einstein believed in that, they knew that science contains infinite mystery). Collective consciousness mystifies "rational science"; many scientists (and nonscientists) believe that rational science will explain everything and there will be no mystery any longer. This is not a delusion, this is a belief; it is not based on any scientific data, it contradicts mathematics (Gödel theorem), which confirms that this is a mystery, a contemporary myth. Dynamic logic gives the mathematical basis for overcoming this quandary. Let me repeat, rational understanding of science and mind contradicts to mystery of symbols connecting conscious and unconscious, which forms the basis of any synthesis. This contradiction prevents formation of contemporary synthesis. Will we overcome this quandary?

Advancement of Western culture during 4000 years proceeded along a razor edge between differentiation and synthesis. Interactions of these opposing factors determined the progress. The balance between them was traditionally supported by art and religion, which reconnected conscious and unconscious. Music played a fundamental role in maintaining this synthesis, like tonal music during the Renaissance. Currently the balance is tilted toward differentiation. However, while immersing into the multiform of life, the collective soul seeks unity in the sublime.

In the midst of desolation and exhaustion of contemporary culture, what could arouse any consoling expectation for the future? Possibly, the only way for future synthesis is individuation—creation of an individual psychological space, where all beginnings and ends belong to an individual human being. A space, where "edges between the soul and body" are mended. In transformation from collective to individual, the inspiration would come from synthesis of scientific and spiritual, of the laws of cultural evolution of beauty, music, and human consciousness.

CHAPTER 8

Musical Emotions and Personality

Contents

Abstract

Music emotion theory in this book explains wide expanses of music practice as well as its millennial mysteries in relation to cognition. Theoretical predictions are confirmed in experiments. Millions of music lovers relate their fascination with music to musical emotions. Still many experts in music from Hanslick to Boulez and Kivy discard emotions as an essential content of music. Why? The chapter relates this controversy to different types of personality, to opposite attitudes of conscious and unconscious, and explains why sometimes people most gifted with understanding emotions discard them as essential for perception of music. I relate personality types to love at first site, divorce, choice of profession, and other important life choices where opposite attitudes of conscious and unconscious may interfere with wise decisions.

A CONTROVERSY IN PERCEPTION OF MUSICAL EMOTION

There is an ongoing controversy in current music cognition research. This book presented theoretical and experimental arguments that many musical emotions are fundamentally different from everyday emotions, and the number of musical emotions is huge; many people feel diverse musical emotions in every musical phrase of every significant composer. Still other distinguished scholars maintain that musical emotions are no different from basic emotions such as sadness, fear, and joy, and their number is no different.

This controversy is paralleled by the fact that some highly achieving and recognized musicians and musical critics including Hanslick, Cage,

Music, Passion, and Cognitive Function.
DOI: http://dx.doi.org/10.1016/B978-0-12-809461-7.00008-X

Boulez, and many others do not value emotional contents of music, search for other musical meanings, and insisted that emotional contents distract from "true" meanings of music. This chapter discusses a counterintuitive conjecture about relations between emotional intelligence and music perception that might explain these controversies.

In the following sections I would like to explore an idea that both opinions about musical emotions might be valid. Some people indeed feel different emotions in every musical phrase, other people do not. This difference is related to the type of personality of the listener, and this relationship is opposite to what could be intuitively expected.

PERSONALITY TYPES AND EMOTION PERCEPTION

Jung (1921) described different types of personality. Here I pay attention to two psychological types related to perception of musical emotions, and accounting for most people: thinking or conceptual type personality and feeling or emotional type. A conceptual personality holds in one's consciousness a variety of conceptual knowledge and easily manipulates it, adequately understands conceptual thoughts of others, freely and adequately uses it in one's life. He or she is not equally conscious about one's emotions. Emotional personality is an "opposite" type; she is conscious about a large number of emotions, can easily manipulate them, can recognize in others, but not equally conscious about conceptual knowledge.

Understanding one's own psychological type as well as those of others is important for many reasons, and yet it is difficult because of complex interaction of conscious and unconscious, which have opposite attitudes. Abilities in one's consciousness come naturally and easily; these are natural abilities that one has. Yet they might be difficult to recognize as one's greatest gift exactly because they come easy, without much effort. On the other hand, what is unconscious is undeveloped, primitive, infantile. But being unconscious these abilities and urges might get to the guts, strongly disturb, and often are perceived as the most important and true self.

For the purpose of this chapter the following counterintuitive properties of personalities are important. A thinking-conceptual type is not much excited by conceptual contradictions in a scientific dispute, he or she would easily consider opposite points of view in his or her consciousness, select appropriate arguments, and would not get provoked into uncontrollable arguments by a new point of view. On the opposite, a person who gets excited by conceptual arguments is typically an emotional type.

Conceptual type being not conscious of a variety of emotions could be deeply touched by emotional arguments, by poetry, or by music. Emotions expressed by music are needed for a conceptual type for internal emotional regulations, and this type may strongly and affectionately respond to music. On the opposite, emotional type is conscious of a variety of emotions. A strongly emotional music does not necessarily bring new emotional knowledge to this personality type. Perception of musical emotions thus could be "opposite" to personality competences: conceptual type may strongly react to musical emotions, whereas emotional type may be indifferent to emotional music.

Conceptual types may prefer classical tonal music, which over centuries has been developed for creating strongly emotional music (Bach, Beethoven, Tchaikovsky), whereas emotional types may prefer atonal music created for eliminating emotions from music. Considering "pure" emotional or conceptual types is a simplification, but this simplification enables one to understand complex intricacies of music perception among many other things important in life.

Related Controversy in Emotion Research

In emotion research there are similar controversies to those in musical emotionality research. Scientists studying emotions take opposite views on the role of conceptual mechanisms in forming emotional experiences. Some emphasize the fact that emotions are phylogenetically ancient mechanisms, humans share emotional experiences with prehuman animals, emotions are essentially preverbal and cannot be adequately expressed in words. Other scientists investigating emotions emphasize that words play an essential role in forming emotional experiences and only after labeling an experience by a specific emotional word the emotion is experienced. Both views are discussed in many reviews.

These opposite points of view could be again due to differences in personality types. A conceptual type, let us repeat, is not fully conscious of his or her emotions, cannot manipulate them at will, to some extent emotions may take hold of a conceptual type. Verbalizing emotions diminishes their intensities and enables their conscious "ownership," which might be important for a conceptual type. On the opposite, an emotional personality type consciously "own" and understand her emotions without interfering categorization processes and labeling them. These attitudes of conceptual personality types are subjectively unconscious and may unconsciously affect experimenters studying the emotions as well as participants in experiments, and results of experiments will reflect the degree of being conscious about one's emotions.

LOVE AT FIRST SIGHT, DIVORCE, AND CHOICE OF PROFESSION

Love at first sight is experienced by many and is a common topic in literature for thousands of years as well as in today's novels and films. Plato explained love as an attraction of two halves of the original androgynies in which a man and a woman were unified. When these halves see each other, they feel an immediate attraction. This myth can be told in accord with contemporary psychology. According to Jung, in most people one personality type predominates in consciousness and is well developed, the opposite type is unconscious and poorly developed. Thus a man or woman are not "complete" individuals, we are more like "halves." When the "halves" see their opposites, they might immediately feel an attraction, the love at first sight.

Everyone can see around that often people are instinctively attracted to the opposite personality type. Why? Forming a couple demands a lot of efforts. Isn't it easier to make a couple when people are similar? Still we are often attracted by the opposite. The opposite personality attracts because it has what one misses. Many couples involve one person who is a conceptual type and another, an emotional type, together they could make one "whole." Men are more often a conceptual type and women are more often an emotional type, but it does not have to be this way. Of course conceptual and emotional types are very approximate, rough characterizations of personalities. There are many books, including Jung, entirely devoted to personality. But this chapter is not about types of personalities, however fascinating this topic might be, this chapter touches personality types to the extent it might help to understand controversies in perception of musical emotions.

To understand differences between conceptual and emotional types, it is important to appreciate the fact that attitudes of unconscious are compensatory and opposite to conscious ones. This causes contradictions between conscious and unconscious in every individual, and these contradictions involve self-perception and self-evaluation, which have misled many outstanding people in understanding themselves and their talents. Misunderstanding oneself and applying one's efforts and talents in a wrong direction could be a grave error. Why people do not understand oneself is poorly studied in science. Helping one to understand oneself is usually treated in self-help books. This is a fascinating and important topic.

For example, a conceptual type easily manipulates various conceptual ideas, this ability is his strongest talent. But exactly because this comes easy, often one may not value it. On the other side, emotions in a conceptual type are mostly unconscious, not well understood, and therefore are infantile and primitive, follow standard clichés. But because they are mostly unconscious, they "get to the guts" and often a conceptual type thinks of his primitive emotions as his most valuable and true self. Similar could be self-misunderstandings of an emotional type, who may underappreciate her diverse and complex emotions that are natural and easy, and overvalue as her true self mostly unconscious, simplistic, and primitive thoughts.

I know of these types of misunderstanding oneself from personal experience. I am a conceptual-thinking psychological type. My greatest successes in life have come from lucky circumstances, which early in my life have led me to make the correct choice to become a physicist; and indeed new scientific ideas come easy to me, and correctly correspond to questions I study. But I have not known that my emotions are primitive typical clichés. Being mostly unconscious, they strongly disturbed me, got to the guts, and for a long time I considered them my most true self.

My wife on the opposite is an emotional type. Her emotions are diverse and appropriate. She experiences them without hesitation and effort and they always correctly correspond to situations. Seeing this everyday I doubted that they are as deep and important as my emotions coming to me with much difficulty and disturbing my inner self. But life proved to me that her emotional reactions have always been right.

After reading much Jung, and analyzing life situations, gradually I have accepted this fact, her emotions are more adequate than mine. And if today some emotions disturb me, e.g., a song does not let me go and continuously repeats in my memory, I turn to my wife. Of course first I think about the reason myself. When I listen to Bach, my emotional and conceptual understanding are in harmony. But it might happen that upon my conceptual analysis, I find the song rather simple and cannot identify any reason why it should strongly affect me. This could be disturbing. If this feeling persists, I would ask for my wife's opinion. If I am emotionally unsettled and do not like it—i.e., I cannot find a conceptual explanation for my feelings—I would insist for her explaining to me what is the reason. Amazingly, she may come with a deep explanation, which eluded myself, still obviously correct. Unwanted, unjustified emotions let go. So indeed conceptual and emotional types may complete each other.

Understanding between partners does not always win over doubts. For the sake of argument, I assume for the next paragraph that a couple is a man of conceptual type and a woman of emotional type. It does not have to be this way, although an opposite case is rare. The reasons for misunderstanding oneself extends to misjudging the other. A conceptual type struggles with his undeveloped, infantile, and unconscious emotions, which might strongly disturb him. When observing his emotional type partner, he may notice that she is not disturbed by emotions, she easily manipulates emotions that come to him with so much difficulty. He might come with conclusions opposite to the reality, he may decide that she is not sincere, that her emotions are easy to her because she is superficial and shallow. Exactly those abilities and character traits of her that initially so much attracted him turn to their opposite.

Similar misunderstandings might occur on the other side of the relationship. Thoughts and conceptual thinking are unconscious in an emotional type, they come with difficulty and might disturb her from the unconscious. Observing how easily her partner manipulates concepts she might come with conclusions opposite to the reality, she may decide that he is not sincere, that his thoughts are superficial and shallow. Exactly those abilities and character traits of him that initially so much attracted her turn to their opposite. These kind of misunderstandings often lead to disappointments and divorce.

The second most important decision everyone must make in life is to identify one's greatest talent and to direct one's energy accordingly. Equally important is to understand one's weaknesses and to avoid channeling energy and life into a wrong direction. One has to correctly select one's profession. Again a controversy between conscious and unconscious attitudes of personality makes the choice difficult. It might be difficult for a talented person to accept that his or her true call in life is to do what comes easy and naturally. Many talented people are seduced by what is challenging and difficult and avoid what is easy and comes naturally, missing their true call. These tragedies of talented people who have not realized their creative potentials are invisible to public, naturally in the focus of public attention are successful people successfully realizing their gifts.

MUSIC AND EMOTIONS

Musical emotions cannot be named in language, we do not have words for them. This book has presented a theory advocating their fundamental role in cognition, experiments tentatively confirming this theory, and their role in evolution of culture. And yet we know that among highly accomplished musicians and musical critics, some deny importance of musical emotions as an essential part of music. Although majority of people listening music for enjoyment testify that their enjoyment comes from musical emotions, many professionals in the area of music deny importance of emotions to music and insist that other properties of music are important, such as musical form.

In this chapter I explore a tentative explanation for this controversy. Subjective perception of music emotions is significantly different among people and related to the type of personality in an unexpected way. People of emotional type, who perceive large number of emotions, easily manipulate them; given a musical talent they could be experts in creating and understanding a huge variety of music emotions, but this type of people could be least interested in emotions: they do not need musical emotions for their emotional life.

The properties of personality types may be a reason for the opposing aesthetic, scientific, and professional views about the importance of music emotions. In my personal experience, observing people I know that it often happens that people of great emotional intelligence are not much excited by musical emotions. They are not necessarily deeply touched by classical tonal music. They could be more interested in atonal music that does not express emotions. They can be more attracted by "pure play" of musical forms. On the opposite, thinking-conceptual personality types could be indifferent to so-called "brainy" atonal music, which they perceive as trivial and boring, they are rather excited by emotional classical music. The opposition between conscious and unconscious attitudes might be the reason why many talented people avoid areas where they could do the greatest contributions to culture.

Possibly the ability for fine discriminations among a manifold of emotions might be independent from the ability to connect them with fundamentally important aspects of cognition and life and to create new emotions. And of course musical abilities are also separate. This is the reason why great composers like Bach, Mozart, Beethoven are rare, even so there are many people with musical abilities, like perfect pitch. I would add that perfect pitch (an ability to exactly hear the tone, the frequency of a string or key) is independent from the ability to perceive fine distinctions of emotions in music. These general hypotheses derived from the theory of personality types could be verified in psychological experiments and could become directions for further experimental research.

It is interesting to note that Schoenberg's dodecaphonic idea that all 12 notes must be played before repeating a key (in order to avoid tonic), may seem an arbitrary and strikingly shallow rule for a person of conceptual type. Despite the rule arbitrariness and conceptual simplicity, atonal music captured a significant part of music compositions in the 20th century. It seems to be of more interest to musicians than to public enjoying music (possibly because a significant part of musicians are of emotional type). It would be interesting to verify experimentally if indeed atonal music is preferred by personalities of emotional type.

NUMBER OF EMOTIONS

How many emotions exist and are musical emotions same or different from usual emotions of joy, sadness, fear, etc.? This book has summarized a theory relating musical emotions to cognitive dissonances; it follows from this theory that there should be a huge number of musical emotions corresponding to a number of cognitive dissonances. The theory is confirmed experimentally, and the role of musical emotions has been traced in evolution of culture. But one important aspect of this theory, existence of a large number of musical emotions has not been experimentally verified. Today there is no reliable way to measure the qualitative value of emotions, to measure the difference between sadness and fear, or between various phrases in, say, Chopin's Nocturne Op. 9 number 2.

Let me repeat, the theory in Chapter 4, Music, relates musical emotions to cognitive dissonances. The number of cognitive dissonances potentially is very large, since every concept may contradict to some extent many other concepts. The number of concepts likely is larger than the number of words, since words and phrases describe concepts. It follows that the number of cognitive dissonances might be in tens or hundreds of thousands, or even millions. And similarly large might be the number of musical emotions. This very large number perceptually could correspond to a continuum of emotions that some people hear in music of every significant composer. Still some musicologists insist that musical emotions are no different than basic emotions, named by emotional words. Differences in personalities might affect perception of emotions, as discussed in this chapter, and be the reason for this controversy.

Let me remind that the theory of musical emotions developed in previous chapters is based on a fundamental difference in emotionality of human voice when used for language and for singing. By relating singing to cognitive dissonances, more exactly to a need for overcoming them, I came to a possibility discussed above that the number of musical emotions might be very large. The number of emotional words in English is much smaller. As discussed in Chapter 1, Theories of Music, and Chapter 2, Mechanisms of the Mind: From Instincts to Beauty, the total number of words designating emotions is about 150; however, the number of different emotions is much smaller. Depending on how "different emotions" are defined these 150 words designate between 2 and 20 different emotions.

And still qualitative measure of musical emotions is possible. A recent publication (Neumann et al., 2015) explores relations between music genres and personality types. This paper discusses that personality types are organized around cognitive dissonances specific for each personality type. And although there are no words specifically designating various cognitive dissonances or musical emotions, that paper has identified correspondence between musical emotions and "bags or words." Bag of words is a model used in text study, which consists of a set of words. That paper has demonstrated that personalities, as expressed in song lyrics, may be related to essays written by the experiment participants. This can be further used to predict participants' personality based on the similarity of their writing style to lyrics of the songs of various music genres, and that songs can be automatically classified into music genres based on the similarity of those song lyrics to essays of participants with known personalities.

Method details *(shortened from Neumann et al., 2015). We used a database of 17,495 songs categorized into music genres and 2468 essays written by participants whose personalities have been assessed through the five factor model of personality. The method uses sophisticated tools of Natural Language Processing.*

The paper uses musical genres identified in (Rentfrow and Gosling, 2003):
-Reflective & Complex (covering blues, jazz, classical, and folk music)
-Intense & Rebellious (rock, alternative, and heavy metal music)
-Upbeat & Conventional (country, soundtracks, religious, and pop music)
-Energetic & Rhythmic (rap/hip-hop, soul/funk, and electronic/dance music).

And personality types, defined by the Big Five personality model:
-Extraversion (E) involves an "energetic approach" to the social and material world and includes traits such as sociability, activity, and positive emotionality.

-Agreeableness (A) involves a pro-social and communal orientation and includes traits such as altruism, tender-mindedness, trust, and modesty.

-Conscientiousness (C) describes socially prescribed impulse control and goal-directed behavior.

-Neuroticism (N) involves negative emotionality and feeling anxious, sad, and tense. This factor is sometimes referred to through its opposite pole: Emotional stability.

-Openness to experience (O) describes the breadth, depth, and originality of the person's mental and experiential life.

Results.

The lyrics of music genres can be used to predict personalities, and conversely, personalities can be used for genre identification and classification.

This paper does not give a direct answer to the question of the multiplicity of musical emotions. The reasons are that the number of musical genres and personality types are limited. Yet an idea of using text bag models to go beyond the limits of emotional words points to a general approach that could be used for identifying emotions of cognitive dissonances and musical emotions. Bag models were used by Petrov to identify the number of different emotional words in English. A similar idea could be applied to analyzing musical "texts," by using one of several available formats. A first step could be to establish how many different musical phrases exist in works of significant composers, the next step would be to identify musical phrases with musical emotions.

CONCLUSION

The function of music in cognition and consciousness, as discussed in Chapter 7, Music and Culture: Parallel Evolution, is fundamental for cognition and cultural evolution. And still there are tremendous differences among people in perception and understanding the role of emotions in music—whereas listeners of music identify their enjoyment of music with musical emotions, many accomplished musicians, including Cage and Boulez, deny a fundamental role of emotions in music. This chapter explored a possible explanation for this controversy by different perception of music by people of different emotional types and opposing attitudes of conscious and unconscious.

It is possible that prevailing intuitions about emotions, especially about aesthetic and musical emotions should be reconsidered. There is a counterintuitive contradiction between dominant abilities and subjective feelings of the "essence of self." One's strongest natural abilities are conscious and come "easy," for these reasons they might be perceived subjectively as "nonessential" and be ignored. On the opposite, one's weakest abilities are nonconscious, from the nonconscious depths of psyche they might excite one's imagination, be perceived as difficult and therefore important, "the essence of true self."

How this hypothesis could be tested? This chapter explores differences in perceiving musical emotions as related to personality types. I looked at fundamental differences related to two Jungian types and musical emotions related to the "Big Five" types of personalities. There are many more personality types than considered in this chapter. The ultimate answer to the controversy about musical emotions will be found when we would learn how to measure qualitative differences among emotions, including musical emotions.

CHAPTER 9

Other Aesthetic Emotions

Contents

Abstract

Near 2,000,000 sites on the web refer to "aesthetics." But definitions of aesthetics, even if given, are circular at best (related to beauty, which is related to aesthetics). Chapter 2, Mechanisms of the Mind: From Instincts to Beauty, gives a scientific definition of aesthetics and the beautiful, which corresponds to a wide understanding, and completes Kantian idea that aesthetics is related to knowledge. Aesthetic emotions measure satisfactions of the knowledge instinct. Near the top of the mental hierarchy they are related to understanding of the highest meaning and purpose, and experienced as the beautiful. This chapter discusses several types of aesthetic emotions, in addition to musical emotions they include emotions of cognitive dissonances, emotions of prosody and their relations to language grammatical inflections and cultures. This book does not touch on aesthetics of poetry, literature, and visual perception.

KANTIAN AESTHETICS

Aesthetic emotions since Kant (1790) are emotions related to knowledge. Kant identified emotions of the beautiful with the understanding of the highest meaning and purpose. He identified the emotions of spiritually sublime with the understanding of actions required to make one's life meaningful and purposeful. The first to use aesthetics as related to spiritual perception was Baumgarten, in "Aesthetics," 1750. Baumgarten thought that aesthetic emotions originate in eyes due to a special perception ability. Cognitive-mathematical theory discussed in this book continues Kant's tradition by identifying aesthetic emotions, including musical emotions with special emotions satisfying the knowledge instinct (KI). It is an extension of Grossberg–Levine theory of drives and emotions to spiritual emotions. The foundational idea of this theory is that

Music, Passion, and Cognitive Function.
DOI: http://dx.doi.org/10.1016/B978-0-12-809461-7.00009-1

emotions are neural signals, states, and related feelings whose functions are related to satisfaction of instinctual drives, and aesthetic emotions are related to the KI.

We discussed experiments, which confirm our theory and ideas of Kant and dis-confirm Baumgarten's ideas.

Usually psychologists study so-called "basic emotions." No agreed upon definition of basic emotions could be found, so for the case of clarity and simplicity I would define basic emotions as related to bodily instincts such as instincts for food, survival, and procreation. We have specific words in language for basic emotions.

MUSICAL EMOTIONS

The main topic of this book are musical emotions whose essential function is to satisfy the KI by helping to overcome contradictions in knowledge. Knowledge has quickly accumulated with evolution of language. This diverse knowledge differentiates consciousness and splits the inner world into pieces. Differentiation is achieved at the cost of synthesis; this threatens the evolution of culture. Musical emotions unify the inner world, produce synthesis and wholeness, and enable personal development and continuation of culture.

EMOTIONS OF COGNITIVE DISSONANCES

Contradictions in knowledge are emotionally unpleasant. These unpleasant emotions are emotions of cognitive dissonances. Why do we have these emotions? What is the reason for their evolution? It is hypothesized that their positive influence on cognition might be a need to make fast decisions, e.g., when a life is threatened, even at the expense of the detailed understanding.

It is known that emotions of cognitive dissonances often lead to an immediate elimination of the source of discomfort, which is a contradictory knowledge. And usually a contradictory knowledge is immediately discarded. Because every knowledge contradicts to some other knowledge, cognitive dissonances could have prevented accumulation of knowledge, and the culture itself. The ability to accumulate knowledge and develop culture requires an ability to overcome cognitive dissonances without discarding knowledge. In this book we concentrated on this negative aspect of cognitive dissonances and an ability to overcome them. It follows from language—cognition interaction model, considered in

Chapter 3, Language and Wholeness of Psyche, that the ability to overcome cognitive dissonances must have originated from emotionality of the voice. This has led to the evolution of music ability.

EMOTIONS OF PROSODY AND CULTURES

Prosody is emotionality of voice. In prehuman animals voice is always emotional. Prehuman animals can vocalize only if they experience emotions. Vocal muscles are controlled by ancient involuntary emotional areas in brains. Emotions and vocalizations therefore are inseparable. Evolution of language required rewiring of the brain. Humans have recent emotional centers in cortex; therefore we can to a significant degree control our vocalizations voluntarily, which is necessary for language.

Involuntary emotionality of voice has been significantly reduced. With evolution of language an ability for strongly emotional voice has mostly evolved into a separate ability for song and music. We can voluntarily or partly so make our voice highly emotional, e.g., during a brawl. But even more important is the remaining weak emotionality of language prosody in everyday "nonemotional" speech. This everyday low-emotional prosody performs a highly important cognitive function: it connects sounds of words with their cognitive meanings. Let me emphasize this possibly nonobvious point. Language and its main way of functioning, speech, can only function if sounds of words are perceived emotionally. If a word sound produce no emotions and no motivations, this word has no meaning. I will dwell on this point because it contradicts accepted understanding. "Emotional speech" often is referred to as a synonym of meaningless or at least devoid of deep meaning, which could be true, especially if emotionality is high and emotions overtake the reason. Here I emphasize the opposite point: usually no emotionality also convey no meaning. Proper emotionality is essential.

Let me make a step back. Unemotional language could be meaningful. When a professor gives a lecture, the best students may receive emotional motivations from understanding the meaningful connections among various parts of knowledge. But this is a special case, and only a small part of students get excited during lecture. In everyday "unemotional" conversations, when listeners do not *have to* understand the meaning of what they hear, they either get subtle motivation from the prosody of speech, or remain indifferent. And most of the time people ignore each other in everyday conversations.

This subtle emotionality of prosody is essential on both sides of speech: a listener needs to perceive emotionally what he hears to perceive any meaning. But, possibly even more important is that a speaker has to experience some degree of emotionality about what he is saying, otherwise his speech may intrinsically be meaningless.

Emotional connections between sounds and meanings may differ among languages. All languages in their evolution have been losing these emotional connections. Ancient "hardwired" connections between emotional brain centers and vocal muscles have been replaced by habitual connections. These connections remain stronger in languages, which sounds change slowly; on the other hand, languages which sounds change historically fast may lose emotional connections between sounds and meaning. Historical changes of language sounds depend on grammar. Sounds of strongly inflectional languages with many affixes, endings, and other inflectional devices change slowly because pronunciations of affixes are controlled by few rules. These few rules are used in every phrase. A child even without schooling, without knowing which grammatical case or gender should be used, learns to pronounce inflections properly. Vocal tract and mouth muscles for pronunciation of affixes (and other inflections) are preserved throughout population and are conserved through generations. Inflections to some extent control pronunciation of words and inflections literary are "tails that wag the dogs," they fix language sounds and therefore emotional prosody and meanings. Humboldt likely meant this by "firmness" of inflectional languages. When inflections disappear, this anchor does not exist, and the sounds of language become fluid and may change with every generation.

In English language this occurred after transition from Middle English to Modern English. Most of the inflections have disappeared and sounds of the language were changing within each generation. English lost excessive emotionality and evolved into a powerful tool of cognition. English spreads democracy, science, and technology around the world. This has been made possible by conceptual differentiation empowered by language. Some professions require people to think, thinking becomes independent from emotional connections between sounds and meanings. The absence of these autonomous, uncontrollable emotions is very useful, e.g., in science and engineering. This might be the reason why English is an international language of science.

But the situation is different in other areas where hard thinking is not professionally required. In these areas people's thoughts have been directed by ancient emotional connections. This preserved traditional values; they are not necessarily "best" but society have survived by relying on traditional ways. Recently traditional values lost much of their power. Majority of people in the United States and in many other European countries are convinced that fast changes happen because people are getting smarter and new social values are necessarily good. But this is not so. It might be useful to realize that some of social changes depend on historically random changes in language prosody.

Current English language cultures face internal crises, uncertainty about meanings and purposes. Many people cannot cope with diversity of life. Future research in psycholinguistics, anthropology, history, historical and comparative linguistics, and cultural studies will examine interactions between languages and cultures. Initial experimental evidence suggests that emotional differences among languages are consistent with this hypothesis.

Sounds in some languages change slowly. Correspondingly emotional connections between sounds and meanings persist over long time. Among most inflectional languages is Arabic, unlike in Indo-European languages, inflections are not only affixes, but vowels in the middle of the word, while the word roots are consonants. Because of this inflectional structure, Arabic preserves in its sounds ancient meanings. Arabic is not flexible nor adaptive for science and engineering, and airplanes are not built in Arabic language. Instead Arabic preserves traditional values. Strong emotional connections between the sounds and meanings make Arabic speakers certain about their life values. This might be a reason why some Arabic people tend to succumb to propaganda more so than people speaking other languages. This contributes to the danger of the current world, and experts in security and international negotiations must be aware of the role of language prosody in today's world.

Possibly music in Arabic cultures has not have to evolve to the extent European classical music has have because emotionality of Arabic language have sustained high level of synthesis.

AESTHETICS OF POETRY, LITERATURE, AND VISUAL PERCEPTION

This book does not touch on these topics. I would like only to emphasize that the understanding of the beautiful discussed in Chapter 2, Mechanisms of the Mind: From Instincts to Beauty, must be a general guiding principle. Emotions of the beautiful are experienced when the contents of the highest representations in the mental hierarchy, which are the highest meaning and purpose, are better understood.

CHAPTER 10

Future Research and Summary

Contents

Abstract

Future research will have to address measure of qualities of musical emotions, e.g., measure of emotional difference between two musical phrases. Similarly, future research will have to measure emotional difference between two cognitive dissonances, as well as emotional differences expressed in prosody. This will create a possibility to measure the number of aesthetic emotions. This chapter discusses possible approaches and expected difficulties.

AESTHETIC EMOTIONS

Measuring Musical Emotions

The fundamental missing experimental ability is how to measure qualities of musical emotions. We do not know how to measure the difference between emotions of any two musical phrases of Chopin, Bach, or Lady Gaga. Therefore laboratory experimentation should be directed at the operational definition and measurements of musical emotions. According to the reviewed theory, the function of musical emotions in cognition is to restore synthesis, when it is damaged by differentiation. Such a condition is similar to cognitive dissonances. The fundamental prediction that music helps overcoming cognitive dissonances have been experimentally confirmed, but the quality of emotions cannot be measured today.

Several approaches can be explored. Well-developed experimental techniques used to study cognitive dissonances can be used to study the proposed role of musical emotions in reconciling contradictions in

Music, Passion, and Cognitive Function.
DOI: http://dx.doi.org/10.1016/B978-0-12-809461-7.00010-8

consciousness. First, various types of cognitive dissonances can be created in subjects using standard techniques. Second, various types of music can be assessed for its efficiency in reconciling specific types of conceptual dissonances. In this way various types of music, which are known to create musical emotions, can be connected to reconciling specific dissonances. Descriptions of cognitive dissonances in words could be scientifically connected to describing musical emotions in words.

The standard approach in psychology to measure similarities—distances among emotions can be used. First, subjective measures of emotional similarities can be collected resulting in a similarity matrix; second, these measures are analyzed mathematically to determine independent emotions, e.g., eigenvectors and eigenvalues. These results can be connected to human feelings by using subjective judgments. Experimental psychologists should expect difficulties related to an expected large number of emotions. Human participants in these experiments cannot be expected to accurately measure millions or even thousands of similarities. Therefore standard techniques should be modified.

Neuroimaging techniques can be used to identify the brain regions involved in musical emotions. Different musical emotions are expected to excite different neural patterns in emotional regions in the brain. The structure of musical emotional space, such as eigenvalues, can be investigated through mathematical methods. Existing mathematical techniques of multidimensional scaling can be used. A word of caution: standard experimental techniques average measurements over many subjects. This is not going to be useful for musical emotions because different participants of these experiments might feel different emotions, when listening to the same music. Even the same subject may experience different emotions when listening to the same music at different times.

Another technique could bring a different degree of "objectivity" to musical emotions. Music can be written as text; a number of techniques can be used today, e.g., mp3. Standard techniques of text analysis can be used to identify (1) records of emotions in musical texts and (2) different musical emotions as manifested in musical texts. Finally, human judgments could be used to relate these "textual emotions" to human emotional feelings.

Measuring musical emotions could be used to measure the number of different emotions in various composers and performers, it would be interesting to compare emotional spaces, e.g., of Eminem and Beethoven to confirm or disprove various expectations. Timbre in music and language might be related to this discussion. Levitin (2006) writes that timbre characterizes individual performers more than any other aspects of music. Patel (2008) suggests that language uses timbre systematically more than music does. Has timbre evolved as "semantic," whereas melody "emotional"? Is harmony related to the mind hierarchy? Are these intuitions just shallow metaphors or meaningful, experimentally testable hypotheses related to the initial separation of voice into language and music, and to further evolution of cultures and consciousness?

Emotions of Cognitive Dissonances

A related research should study emotional spaces of cognitive dissonances. Again well-explored standard psychological techniques of studying emotions (Fontanari et al., 2012) can be used. First, subjective measures of emotional similarities can be collected; second, these measures are analyzed mathematically to determine the number of independent emotions, e.g., eigenvalues. Experimental psychologists again should expect difficulties related to an expected large number of emotions. Another specific difficulty is to make sure that subjects concentrate on emotions of cognitive dissonance and are not distracted by semantics of phrases used to create the dissonances.

Other experimental techniques discussed for musical emotions can be used to measure emotions of cognitive dissonances. Every cognitive dissonance is expected to create a pattern of neuronal excitations in emotional brain centers. These patterns can be analyzed by using standard mathematical techniques present in every data analyzes package, such as eigenvalues or multidimensional scaling. Again, these mathematically identified emotions can be related to human feelings using subjective judgments. Thus cognitive dissonances could connect a wealth of aesthetic emotions that have never been studied to language description of these emotions by phrases used to create cognitive dissonances.

Emotions of Prosody

Nonintentional emotions of prosody that are inherent to language have not been studied. Studies of prosody have concentrated on expressions of basic emotions, e.g., an actor intentionally expresses a particular, well-understood basic emotion and the purpose of the experiment is to see how well participants recognize the emotion. I emphasize here a need to study prosodial emotions that are *inherent to language*, mostly unconscious, and unrelated to expressing some well-understood emotions such as fear, sadness, or joy.

Chapter 8, Musical Emotions and Personality, discussed that prosodial emotions inherent to language have a fundamental cognitive function of connecting word sounds to their meanings. First, this hypothesis should be tested. If confirmed, this will open unlimited directions for future research. How many prosodial emotions are there? To which extent cultural connections between sounds and meanings influence individual understanding? Jung discussed that emotionality and conceptual thinking are opposite abilities; emotions disrupts conceptual thinking, while conceptual thinking disrupts emotionality. Surprisingly little experimental research has been devoted to this fundamental aspect of psychology.

Possibly this research can be added by understanding that interactions between emotional analysis and conceptual thinking reside in emotional prosody of language. While emotional prosody is necessary for connecting sounds of words to their meanings, which is the essence of language, at the same time it interrupts conceptual arguments, which is the function of language in thinking.

This future research could be conducted by considering people of different personality types: emotional-feeling types and conceptual-thinking types. One research direction could be to identify the number of prosodial emotions used by different personality types. Another directions could address specific differences in emotional and conceptual understanding in different types.

A parallel direction, instead of personality types, could explore differences in emotional prosody among languages. What are the differences among languages? Chapter 8, Musical Emotions and Personality, discussed advantages of low-prosodial emotionality of languages (such as English) for differentiation in science and engineering versus disadvantages for synthesis, for understanding fundamental cultural questions, such as the meaning of life—these hypotheses should be tested. Languages with high-prosodial emotionality (Arabic) may have opposite properties: a disadvantage for conceptual differentiation in science and engineering, while an advantage for synthesis in understanding the fundamental life values, including the meaning of life. Emotionality versus conceptuality could be studied for the huge variety of existent languages.

Another research direction can study possible relations between language grammar and emotionality of prosody. Does grammar indeed affects emotionality of prosody? Influence of language properties on emotionality of prosody, cultures, and their evolution opens a wide field for experimental studies. Possibly we would come with novel appreciation of diversity among human cultures.

Music and Culture

Chapter 7, Music and Culture: Parallel Evolution, explored parallel evolution of music, consciousness, and culture. Every example of parallel changes in consciousness and music discussed in the chapter can be considered as a hypothesis requiring experimental tests. Connections between music and consciousness could be explored in individual lives of composers. Connections between music and culture evolution should be explored in a variety of cultures beyond few examples discussed in Chapter 7, Music and Culture: Parallel Evolution. This should be turned into interdisciplinary research attracting linguists, anthropologists, cultural historians, and mathematicians.

Language and Cognition

The theory of musical function in cognition developed in this book takes its roots in mechanisms of interacting cognition and language, the dual hierarchy model discussed in Chapter 3, Language and Wholeness of Psyche. This model contradicts Chomskyan's prediction that cognition and language are independent abilities. Can this model be tested? The model described in this book makes a number of specific predictions, some of them have been experimentally tested and other could be tested in future. Future research should test that language representations are crisper and more conscious than cognitive representations. If cognition and language to some extent are different abilities, can each ability be independently measured? How are people distributed along cognition–language axis? fMRI can be used to identify details of neural mechanisms of each ability and their integration. Can it be demonstrated that inborn language and cognitive representations both are placeholders without concrete contents? Are connections between these representations inborn? Can future research demonstrate that cognitive representations are learned guided by language representations? Is syntax learned guided by relations among objects?

The Beauty and Propaganda

Chapter 6, Experimental Tests of the Theory: Beauty and Meaning, described experimental confirmations of the theoretical prediction that beauty is related not to sex, and even not to a form of art object, but to what creates the meaning and purpose in one's life. Future research should expand experimental and theoretical studies toward narratives, poetry, and visual arts. This research is important for high art, which in the past century have been subverted to concentrating on the *form* of art, as well as to propagandistic material, which becomes a new danger to the world peace and requires understanding of counter measures. In the last century everything related to the meaning and purpose was often assigned to propaganda, this tended to make art minimalistic, meaningless. A new, more sophisticated understanding developed in this book is required to be further developed.

SUMMARY

Why listening to music might be a source of pleasure? Why does music has such a power over us? Two thousand four hundred years ago Aristotle asked this question: why music being just sounds reminds states of human soul? Great thinkers attempted to answer this question for two and a half millennia. The answer was not found. Why such a strange ability to enjoy sounds could emerge in evolution? Does music has a cognitive function? In 1871 Darwin wrote that music "must be ranked amongst the most mysterious (abilities) with which (man) is endowed."

This book offers an answer to this question, a scientific answer connected to Ecclesiastes "with much wisdom comes much sorrow": evolution of language results in fast growth of knowledge. Knowledge leads to contradictions, cognitive dissonances. Human inner world is split; synthesis, a unity of self might be lost. This is unpleasant even painful. If these unpleasant emotions would not be overcome, knowledge would not accumulate, culture would not develop. Music evolves to overcome these discomforts caused by cognitive dissonances and split soul, so that a human soul could unify and culture could continue evolving.

This theory answers difficult questions about music, mysteries which remained misunderstood for millennia. Why does music has such a power over us? Why many people enjoy listening to sad music, which makes us cry? The answers offered in this book are that contradictions in knowledge, cognitive dissonances cause us pain, we live in a sea of grief; these real sufferings in life are much worth than negative emotions experienced in sad music, music helps us to overcome painful emotions unavoidable in life.

A cognitive-mathematical theory offering this explanation is the only theory describing a wealth of empirical data and the only theory making experimentally verifiable predictions. This theory transformed psychology from "soft" science into "hard" science; a new area of science, physics of the mind. Theoretical predictions about the role of music in overcoming cognitive dissonances without discarding knowledge, about unifying the split soul, about relations of beautiful to meaning, and other theoretical predictions have been experimentally confirmed.

This book reviews the first step in identifying the fundamental role of musical emotions in cognition and cultural evolution. Possibly it will contribute to a foundation for a unified field of multidisciplinary study. In conclusion, I would like to repeat that music is the most mysterious of human abilities, appealing directly to our primordial emotions, while connecting them to language and cognition.

LITERATURE

Akerlof, G. A., & Dickens, W. T. (2005). The economic consequences of cognitive dissonance. In G. A. Akerlof (Ed.), *Explorations in pragmatic economics*. New York, NY: Oxford University Press.

Albright, D. (2001). *Untwisting the serpent: Modernism in music, literature, and other arts*. Chicago, IL: Univ. Chicago Press.

Alfnes, F., Yue, C., & Jensen, H. H. (2010). Cognitive dissonance as a means of reducing hypothetical bias. *European Review of Agricultural Economics*, *37*(2), 147–163.

Alper, P. (Ed.), (1994). *Musical worlds, new directions in the philosophy of music*. University Park, PA: Penn State Univ. Press.

Andersson, M. (1994). *Sexual selection*. Princeton, NJ: Princeton Univ. Press.

D'Aquili, E. G., Laughlin, C. D., & McManus, J. (1979). *The spectrum of ritual: A biogenetic structural analysis*. New York: Columbia University Press.

Aranovsky, M. (1998). *Musical text*. Moscow, Russia: National Inst. Study Arts.

Aristotle (1995). In J. Barnes (Ed.), *The complete works:The revised Oxford translation*. Princeton, NJ: Princeton Univ. Press, Original work VI BCE.

Aronson, E., & Carlsmith, M. (1963). Effect of the severity of threat on the devaluation of forbidden behavior. *The Journal of Abnormal and Social Psychology*, *66*(6), 584–588. Available from http://dx.doi.org/10.1037/h0039901.

Arnheim, R. (1969). *Visual thinking*. Berkeley, CA: University of California Press, ISBN 978-0-520-24226-5.

Atran, S., Sheikh, H., & Gomez, A. (2014). Devoted actors sacrifice for close comrades and sacred cause. *Proceedings of the National Academy of Sciences*, *111*(50), 17702–17703. Available from http://dx.doi.org/10.1073/pnas.1420474111.

Augustine, St., see Weiss, P., & Taruskin, R. (1984).

Balaskó, M., & Cabanac, M. (1998). Grammatical choice and affective experience in a second-language test. *Neuropsychobiology*, *37*, 205–210.

Ball, P. (2008). Facing the music. *Nature*, *453*, 160–162.

Bar, M., Kassam, K. S., Ghuman, A. S., Boshyan, J., Schmid, A. M., Dale, A. M., et al. (2006). Top-down facilitation of visual recognition. *Proceedings of the National Academy of Sciences of the United States of America*, *103*, 449–454.

Barbuto, J. E., Jr. (1997). A critique of the Myers-Briggs type indicator and its operationalization of Carl Jung's psychological types. *Psychological Reports*, *80*, 611–625.

Barrett, L. F. (2006). Solving emotion paradox: Categorization and the experience of emotion. *Personality and Social Psychology Review*, *10*(1), 20–46. Available from http://dx.doi.org/10.1207/s15327957pspr1001_2.

Baumgarten, A.B. (1750). Aesthetica. I.C. Kleyb.

Benedek, M., & Kaernbach, C. (2011). Physiological correlates and emotional specificity of human piloerection. *Biological Psychology*, *86*(3), 320–329. Available from http://dx.doi.org/10.1016/j.biopsycho.2010.12.012.

Berto, R. (2005). Exposure to restorative environments helps restore attentional capacity. *Journal of Environmental Psychology*, *25*(3), 249–259.

Berto, R. (2014). The role of nature in coping with psycho-physiological stress: A literature review on restorativeness. *Behavioral Science (Basel)*, *4*(4), 394–409.

Blood, A. J., & Zatorre, R. J. (2001). Intensely pleasurable responses to music correlate with activity in brain regions implicated in reward and emotion. *Proceedings of the National Academy of Sciences of the United States of America*, *98*(20), 11818–11823.

Boethius, 5c., see Weiss, P., & Taruskin, R. (1984).

Brighton, H., Kirby, S., & Smith, K. (2005). Language as an evolutionary system. *Physics of Life Reviews*, *2*, 177–226. Available from http://dx.doi.org/10.1016/j.plrev.2005.06.001.

Brodsky, J. (1989). *North of Delphi, tr. 1991 by the author.* New York, NY: Farrar, Straus and Giroux, 2000.

Brodsky, J. (1989). Поэзия как форма сопротивления реальности, Сочинения Иосифа Вродского, т.VII, оbщ. ред. Я. А. Гордина, Пушкинский фонд, Санкт Петерbург, 2001 [Poetry as a form of resistance to reality, Works of Joseph Brodsky, v.7, J. Gordin (Ed.), Pushkin Foundation, St. Peterburg, 2001] (in Russian).

Brodsky, J. (1991). "Наглая проповедь идеализма", in Иосиф Вродский, "Вольшая книга интервью." Изд. 2, Захаров, Москва, 2000, стр. 512 [Interview with D. Betea. In J. Brodsky, C. L. Haven, and R. Avedon, 2003, Joseph Brodsky: Conversations, University Press of Mississippi.] (in Russian).

Brown, S. (2000). Evolutionary models of music: From sexual selection to group selectionIn F. Tonneau, & N. S. Thompson (Eds.), *Perspectives in Ethology* (Vol. 13, pp. 231–281). New York: Kluwer Academic/Plenum Publishers.

Buchanan, T. W., Lutz, K., Mirzazade, S., Specht, K., Shah, N. J., Zilles, K., et al. (2000). Recognition of emotional prosody and verbal components of spoken language: An fMRI study. *Cognitive Brain Research*, *9*, 227–238.

Cabanac, A., Perlovsky, L. I., Bonniot-Cabanac, M.-C., & Cabanac, M. (2013). Music and academic performance. *Behavioural Brain Research*, *256*, 257–260.

Cabanac, M. (2002). What is emotion? *Behavioural Processes*, *60*, 69–84.

Cage, J. (1951). Music of changes. <https://www.youtube.com/watch?v=B_8-B2rNw7s> (18.04.16).

Cangelosi, A., Bugmann, G., & Borisyuk, R. (Eds.), (2005). Modelling language, cognition and action. *Proceedings of the 9th neural computation and psychology workshop* Singapore: World Scientific.

Cangelosi, A., Greco, A., & Harnad, S. (2000). From robotic toil to symbolic theft: Grounding transfer from entry-level to higher-level categories. *Connection Science*, *12*, 143–162.

Cangelosi, A., & Parisi, D. (Eds.), (2002). *Simulating the evolution of language.* London: Springer.

Cangelosi, A., & Riga, T. (2006). An embodied model for sensorimotor grounding and grounding transfer: Experiments with epigenetic robots. *Cognitive Science*, *30*, 673–689.

Cangelosi, A., Tikhanoff, V., Fontanari, J. F., & Hourdakis, E. (2007). Integrating language and cognition: A cognitive robotics approach. *IEEE Computational Intelligence Magazine*, *23*, 65–70.

Chater, N. (1999). The search for simplicity: A fundamental cognitive principle? *The Quarterly Journal of Experimental Psychology Section A*, *52*(2), 273–302.

Chomsky, N. (1957). *Syntactic structures*. Haag, Netherlands: Mouton.

Christiansen, M., & Kirby, S. (2003). *Language evolution.* Oxford, United Kingdom: Oxford University Press.

Cimprich, B., & Ronis, D. L. (2003). An environmental intervention to restore attention in women with newly diagnosed breast cancer. *Cancer Nursing*, *26*(4), 284–292, quiz 293–294.

Conard, N. J., Malina, M., & Münzel, S. C. (2009). New flutes document the earliest musical tradition in southwestern Germany. *Nature*, *460*, 737–740. Available from http://dx.doi.org/10.1038/nature08169.

Confucius. (551–479 B.C.E. (2000). *Analects* (D.C. Lau, Trans.). Hong Kong, China: The Chinese University Press.

Cooper, J. (2007). *Cognitive dissonance: 50 years of a classic theory.* Los Angeles, CA: Sage.

Cooper, J. (2016). <www.lvbeethoven.com/Oeuvres_Presentation/Presentation-Piano Sonatas-30And32.html>.
Corrigall, K. A., Schellenberg, E. G., & Misura, N. M. (2013). Music training, cognition, and personality. *Frontiers in Psychology, 4*(2), 222.
Coutinho, E., & Cangelosi, A. (2009). The use of spatio-temporal connectionist models in psychological studies of musical emotions. *Music Perception, 271*, 1–15.
Coventry, K. R., Lynott, L., Cangelosi, A., Monrouxe, L., Joyce, D., & Richardson, D. C. (2010). Spatial language, visual attention, and perceptual simulation. *Brain and Language, 112*(3), 202–213.
Craig, D. (2005). An exploratory study of physiological changes during "chills" induced bymusic. *Musicae Scientiae, IX*(2), 273–287.
Cross, I. (2008a). The evolutionary nature of musical meaning. *Musicae Scientiae*, 179–200.
Cross, I. (2008b). Musicality and the human capacity for culture [Special issue]. *Musicae Scientiae*, 147–167.
Cross, I. (2014). Music and communication in music psychology. *Psychology of Music, 42* (6), 809–819. Available from http://dx.doi.org/10.1177/0305735614543968.
Cross, I., & Morley, I. (2008). The evolution of music: Theories, definitions and the nature of the evidence. In S. Malloch, & C. Trevarthen (Eds.), *Communicative musicality* (pp. 61–82). Oxford: Oxford University Press.
Damasio, A. (2001). Music and emotion. Course in Ohio State University. <http://csml.som.ohio-state.edu/Music829D/music829D.html> (29.11.15).
D'Aquili, E. G., Laughlin, C. D., & McManus, J. (1979). *The spectrum of ritual : A biogenetic structural analysis*. New York, NY: Columbia University Press.
Darwin, C. R. (1871). *The descent of man, and selection in relation to sex* (p. 880). London, GB: John Murray.
Davis, P. J., Zhang, S. P., Winkworth, A., & Bandler, R. (1996). Neural control of vocalization: Respiratory and emotional influences. *Journal of Voice, 10*, 23–38.
Day, H.I. (1969). *An instrument for the measurement of intrinsic motivation. Technical report*. An interim report to the Department of Manpower and Immigration, Australia.
Deacon, T. (1989). The neural circuitry underlying primate calls and human language. *Human Evolution Journal, 4*(5), 367–401.
Descartes, R. (1646). *Passions of the soul* (p. 181). Cambridge: Hackett Publishing Company.
Diamond, J. (1997). *Guns, germs, and steel: The fates of human societies*. New York, NY: W.W. Norton, and Co.
Dirac, P. A. M. (1982). *The principles of quantum mechanics* (p. 83). Oxford UK: Oxford University Press.
Dissanayake, E. (2000). Antecedents of the temporal arts in early mother-infant interactions. In N. Wallin, B. Merker, & S. Brow (Eds.), *The origins of music* (pp. 389–407). Cambridge, MA: MIT Press.
Dissanayake, E. (2008). If music is the food of love, what about survival and reproductive success? [Special issue]. *Musicae Scientiae*, 169–195.
Dorak, M.T. (2008). *Music Home Page*. <http://www.dorak.info/music/contents.html>.
Editorial (2008). Bountiful noise. *Nature, 453*, 134.
Ekman, P. (1999). Basic emotions. In T. Dalgleish, & M. Power (Eds.), *Handbook of cognition and emotion*. Sussex, UK: John Wiley and Sons.
Eliot, T.S. (1919). Tradition and the individual talent. In *The Sacred Wood*. Online Source: <http://www.bartleby.com/200/sw4.html>.
Festinger, L. (1957a). *A theory of cognitive dissonance*. Evanston, IL: Row, Peterson.
Festinger, L. (1957b). *A theory of cognitive dissonance*. Stanford, CA: Stanford University Press.

Festinger, L. (1962). Cognitive dissonance. *Scientific American, 207*, 93–102.

Festinger, L., & Carlsmith, J. M. (1959). Cognitive consequences of forced compliance. *Journal of Abnormal Psychology, 58*(2), 203–210.

Festinger, L., & Hutte, H. A. (1954). An experimental investigation of the effect of unstable interpersonal relations in a group. *Journal of Abnormal Psychology, 49*(1, Part 1), 513–522.

Fitch, W. T. (2004). On the biology and evolution of music. *Music Perception, 24*, 85–88.

Fontanari, J. F., Bonniot-Cabanac, M.-C., Cabanac, M., & Perlovsky, L. I. (2012). A structural model of emotions of cognitive dissonances. *Neural Networks, 32*, 57–64.

Fontanari, J. F., & Perlovsky, L. I. (2007). Evolving compositionality in evolutionary language games. *IEEE Transactions on Evolutionary Computations, 11*(6), 758–769, on-line doi:10.1109/TEVC.2007.892763.

Fontanari, J. F., & Perlovsky, L. I. (2008a). How language can help discrimination in the Neural Modelling Fields framework. *Neural Networks, 21*(2-3), 250–256.

Fontanari, J. F., & Perlovsky, L. I. (2008b). A game theoretical approach to the evolution of structured communication codes. *Theory in Biosciences, 127*, 205–214.

Fontanari, J. F., Tikhanoff, V., Cangelosi, A., Ilin, R., & Perlovsky, L. I. (2009). Cross-situational learning of object–word mapping using Neural Modelling Fields. *Neural Networks, 22*(5–6), 579–585. Available from http://dx.doi.org/10.1007/s12064-008-0024-1.

Franklin, A., Drivonikou, G. V., Bevis, L., Davie, I. R. L., Kay, P., & Regier, T. (2008). Categorical perception of color is lateralized to the right hemisphere in infants, but to the left hemisphere in adults. *Proceedings of the National Academy of Sciences, 105*(9), 3221–3225.

Freytag, G. (1897). *Die technik des dramas. Gesammelte Werke* (Vol. 14). Leipzig: S. Hirzel.

Frijda, N. H. (1986). *The emotions*. Cambridge: Cambridge University Press.

Gasset, O. J. (1925). *The dehumanization of art*. Princeton, NJ: Princeton Univ. Press, see 1968.

Gerrig, R. J. (1993). *Experiencing narrative worlds: On the psychological activities of reading*. New Haven, CT: Yale University Press.

Goldstein, A. (1980). Thrills in response to music and other stimuli. *Physiological Psychology, 8*, 126–129.

Golomb, U. (1998). *Modernism, rhetoric and (de-)personalisation in the early music movement*. <http://homepages.kdsi.net/~sherman/golomb1.htm>.

Green, M. C., & Brock, T. C. (2000). The role of transportation in the persuasiveness of public narratives. *Journal of Personality and Social Psychology, 79*, 701–721. pmid:11079236. Available from http://dx.doi.org/10.1037//0022-3514.79.5.701.

Greenberg, C. (1999). *Homemade esthetics: Observations on art and taste*. Oxford: Oxford Univ. Press.

Grewe, O., Kopiez, R., & Altenmüller, E. (2009). Chills as an indicator of individual emotional peaks. *Annals of the New York Academy of Sciences, 1169*, 351–354.

Grewe, O., Nagel, F., Kopiez, R., & Altenmüller, E. (2005). How does music arouse "chills" ? investigating strong emotions, combining psychological, physiological, and psychoacoustical methods. *Annals of the New York Academy of Sciences, 1060*, 446–449.

Groceo, J. (14 c.). (1984). In P. Weiss, R. Taruskin (Eds.), *Music in the western world* (p. 63). New York, NY: Schirmer/Macmillan.

Grodal, T. K. (1997). *Moving pictures: A new theory of film genres, feelings, and cognition*. Oxford: Clarendon Press.

Grossberg, S. (1982). *Studies of mind and brain. D.* Dordrecht, Holland: Reidel Publishing Co.

Grossberg, S. (2014). From brain synapses to systems for learning and memory: Object recognition, spatial navigation, timed conditioning, and movement control.

Brain Research, 1621, 270–293. Available from http://dx.doi.org/10.1016/j.brainres.2014.11.018.

Grossberg, S., & Levine, D. S. (1987). Neural dynamics of attentionally modulated Pavlovian conditioning: Blocking, inter-stimulus interval, and secondary reinforcement. *Applied Optics, 26*, 5015–5030.

Guttfreund, D. G. (1990). Effects of language usage on the emotional experience of Spanish-English and English-Spanish bilinguals. *Journal of Consulting and Clinical Psychology, 58*, 604–607.

Haidt, J. (2000). The positive emotion of elevation. *Prevention & Treatment, 3*, Article 3, 2000.

Hamann, J. G. (2001/1758). *Aesthetica in nuce. Métacritique du purisme de la raison pure et autres textes*. Paris: Vrin, ISBN 978-2-7116-1475-2.

Harrison, L., & Loui, P. (2014). Thrills, chills, frissons, and skin orgasms: Toward an integrative model of transcendent psychophysiological experiences in music. *Frontiers in Psychology, 5*, 790.

Hegel, G. W. (1830). *Hegel's logic: Being part one of the encyclopaedia of the philosophical sciences* (p. 1975). Oxford, UK: Clarendon Press.

Hanslick, E. (1854/1986). *The beautiful in music*. Cambridge, MA: Hackett Publishing Company.

Harmon-Jones, E., Amodio, D. M., & Harmon-Jones, C. (2009). Action-based model of dissonance: A review, integration, and expansion of conceptions of cognitive conflictIn M. P. Zanna (Ed.), *Advances in experimental social psychology* (Vol. 41, pp. 119–166). Burlington, CA: Academic Press.

Harris, C. L., Ayçiçegi, A., & Gleason, J. B. (2003). Taboo words and reprimands elicit greater autonomic reactivity in a first language than in a second language. *Applied Psycholinguistics, 24*, 561–579.

Harrison, L., & Loui, P. (2014). Thrills, chills, frissons, and skin orgasms: Toward an integrative model of transcendent psychophysiological experiences in music. *Frontiers in Psychology, 5*, 790.

Helmholtz, H. L. F. (1863/1954). *On the sensations of tone—As a physiological basis for the theory of music*. New York, NY: Dover Publication.

Hesse, H. (1943). *The Glass Bead Game (Magister Ludi)*. (R.C. Winstons, Trans.). USA: Picador.

Hoeschele, M., Merchant, H., Kikuchi, Y., Hattori, Y., & ten Cate, C. (2015). Searching for the origins of musicality across species. *Philosophical Transactions of the Royal Society B: Biological Sciences, 370*, 20140094. Available from http://dx.doi.org/10.1098/rstb.2014.0094.

Hoffmann, E.T.A. (1813). *Musikalische Novellen und Aufsätze* (Schwaegermann, Trans.). Leipzig: Insel-Bücherei. <http://www.geocities.com/Vienna/Strasse/3732/hoffm-lvb-instumentalmusik.html>.

Honing, H. (2013). *Musical cognition: A science of listening*. New Brunswick, NJ: Transaction Publishers.

Honing, H., ten Cate, C., Peretz, I., & Trehub, S. E. (2015). Without it no music: Cognition, biology and evolution of musicality. *Philosophical Transactions of the Royal Society B: Biological Sciences, 370*, 1664. <http://rstb.royalsocietypublishing.org/content/370/1664/20140088.full>.

von Humboldt, W. (1836/1967). Über die Verschiedenheit des menschlichen Sprachbaues und ihren Einfluss auf die geistige Entwickelung des Menschengeschlechts. Berlin: F. Dummler. In W. P. Lehmann (Ed.), *A reader in nineteenth century historical Indo-European linguistics*. Bloomington, IN: Indiana University Press.

Hume, D. (1748). *Philosophical essays concerning human understanding* (1 ed) London: A. Milla.

Huron, D. (1999). *Ernest Bloch lectures*. Berkeley, CA: University of California Press.

Huron, D. (2006). *Sweet anticipation: Music and the psychology of expectation.* Cambridge, MA: The MIT Press.
Huron, D. (2011). Why is sad music pleasurable?: A possible role for prolactin. *Music Sci, 15*, 146–158.
Ilin, R., & Perlovsky, L. I. (2010). Cognitively inspired neural network for recognition of situations. *International Journal of Natural Computing Research, 11*, 36–55.
Izard, C. E. (1992). Basic emotions, relations among emotions, and emotion-cognition relations. *Psychological Review, 99*, 561–565.
James, J. (1995). *The music of the spheres: Music, science, and the natural order of the universe.* New York, NY: Springer.
Jarcho, J. M., Berkman, E. T., & Lieberman, M. D. (2011). The neural basis of rationalization: Cognitive dissonance reduction during decision-making. *Social Cognitive and Affective Neuroscience, 6*(4), 460–467.
Jaynes, J. (1976). *The origin of consciousness in the breakdown of the bicameral mind.* Boston, MA: Houghton Mifflin Co.
Jepma, M., Verdonschot, R. G., van Steenbergen, H., Rombouts, S. A., & Nieuwenhuis, S. (2012). Neural mechanisms underlying the induction and relief of perceptual curiosity. *Frontiers in Behavioral Neuroscience, 6*, 5. Available from http://dx.doi.org/10.3389/fnbeh.2012.00005.
John, O. P., & Srivastava, S. (1999). The Big Five trait taxonomy: History, measurement, and theoretical perspectives. In L. A. Pervin, & O. P. John (Eds.), *Handbook of personality: Theory and research* (2nd ed., pp. 102–138). New York, NY: Guilford Press.
Joyce, J. (1922). *Ulysses, Vintage Reissue edition, 1990.* New York, NY: Random House.
Jung, C. G. (1921). *Psychological types. Collected works, Bollingen Series XX* (Vol. 6). Princeton, NJ: Princeton University Press.
Jung C.G. (1934). Ulysses. Wirklishkeit der Seele, Rascher, Zurich; 1934. Translation mostly follows Ulysses: A Monologue. In C.G. Jung (Ed.), *The spirit in man, art, and literature* (Vol. 15) (R.F.C. Hull, Trans.) (pp. 109–110, 115–119). Princeton, NJ: Princeton Univ. Press.
Jung, C. G. (1948). *The self, in Aion. Collected works, Bollingen Series XX* (Vol. 9, pp. 23–25). Princeton, NJ: Princeton University Press.
Juslin, P. N., & Sloboda, J. A. (2001). *Music and emotion: Theory and research.* Oxford, GB: Oxford University Press.
Juslin, P. N., & Västfjäll, D. (2008). Emotional responses to music: The need to consider underlying mechanisms. *Behavioural and Brain Sciences, 31*, 559–575.
Justus, T., & Hutsler, J. J. (2005). Fundamental issues in the evolutionary psychology of music: Assessing innateness and domain specificity. *Music Perception, 23*, 1–27.
Kang, M. J., Hsu, M., Krajbich, I. M., Loewenstein, G., McClure, S., Wang, J., & Camerer, C. F. (2009). The wick in the candle of learning epistemic curiosity activates reward circuitry and enhances memory. *Psychological Science, 20*(8), 963–973. Available from http://dx.doi.org/10.1111/j.1467-9280.2009.02402.x.
Kant, I. (1790). The critique of judgment (J.H. Bernard, Trans.). Amherst, NY: Prometheus Books.
Kapitza, S. P. (1994). The impact of the demographic transition. In K. Schwab (Ed.), *Overcoming indifference. Ten key challenges in today's world: World economic forum.* NY: N.Y. Univ. press.
Kapitza, S. P. (1996). Population: Past and future. A mathematical model of the world population system. *Science Spectra, 2*(4).
Kaplan, S. (1995). The restorative benefits of nature: Toward an integrative framework. *Journal of Environmental Psychology, 15*, 169–182.
Kaplan, R. (2001). The nature of the view from home: Psychological benefits. *Environment and Behavior, 33*, 507–542.

Kashdan, T. B., Barrett, L. F., & McKnight, P. E. (2014). Unpacking emotion differentiation: Transforming unpleasant experience by perceiving distinctions in negativity. *Current Directions in Psychological Science, 24*, 10–16. Available from http://dx.doi.org/10.1177/0963721414550708.

Keltner, D., & Haidt, J. (2010). Approaching awe, a moral, spiritual, and aesthetic emotion. *Cognition and Emotion, 17*(2), 297–314.

Kivy, P. (1994). Armistice, but no surrender: Davies on Kivy. *Journal of Aesthetics and Art Criticism, 52*(2), 236–237.

Kivy, P. (2011). *Antithetical arts: On the ancient quarrel between literature and music.* New York, NY: Oxford Univ. Press.

Koelsch, S., Jacobs, A., Menninghaus, W., Liebal, K., Klann-Delius, G., von Scheve, C., & Gebauer, G. (2015). The quartet theory of human emotions: An integrative and neurofunctional model. *Physics of Life Reviews, 13*, 1–27.

Konecni, V., Wanic, R., & Brown, A. (2007). Emotional and aesthetic antecedents and consequences of music-induced thrills. *American Journal of Psychology, Vol. 120* (No. 4), 619–643.

Kosslyn, S. M. (1994). *Image and Brain: The Resolution of the Imagery Debate.* Cambridge, MA: MIT Press.

Hans, K., & Shulamith, K. (1972). *Psychology of the arts.* Durham NC: Duke University Press, Duke University Press.

Kuhn, T. S. (1962). *The Structure of Scientific Revolutions* (3rd ed., p. 1996). See University of Chicago Press.

Lao-Tzu. (6th B.C.E./1979). *Tao Te Ching* (D.C. Lau, Trans.). New York, NY: Penguin Books.

Larson, C. R. (1991). Activity of PAG neurons during conditioned vocalization in the macaque monkey. In A. Depaulis, & R. Bandler (Eds.), *The midbrain periaqueductal gray matter* (pp. 23–40). New York, NY: Plenum Press.

Lavandier, Y. (2005). *Writing drama; a comprehensive guide for playwrights and scriptwriters paperback.* Paris: Le Clown & L'enfant.

Laughlin, Charled D., & d'Aquili, Eugene (1974). *Biogenetic structuralism.* New York, NY: Columbia University Press.

Lebrecht, N. (1992). *The Independent Magazine.* <http://www.music.gla.ac.uk/~tfowler/articles/Boulez.html>.

Lerer, S. (2007). *Inventing English.* Chichester, NY: Columbia University Press.

Levine, D., & Leven, S. J. (Eds.), (1992). *Motivation, emotion, and goal direction in neural networks* Hillsdale, NJ: Erlbaum.

Levine, D. S. (2012). I think therefore I feel: Possible neural mechanisms for knowledge-based pleasure. *Proceedings of IJCNN, 2012*, 363–367.

Levine, D. S., & Perlovsky, L. I. (2008a). Neuroscientific insights on Biblical myths. Simplifying heuristics versus careful thinking: Scientific analysis of millennial spiritual issues. *Zygon, Journal of Science and Religion, 43*(4), 797–821.

Levine, D. S., & Perlovsky, L. I. (2008b). *A network model of rational versus irrational choices on a probability maximization task.* Hong Kong, China: World Congress on Computational Intelligence WCCI.

Lévi-Strauss, C. (1963). *Structural anthropology* (C. Jacobson, B. Grundfest Schoepf, Trans.). New York, NY: Basic Books.

Levitin, D. J. (2006). *This is your brain on music: The science of a human obsession.* London: Dutton.

Levitin, D. J. (2008). *The world in six songs.* London: Dutton.

Lieberman, M.D. (2011). Why symbolic processing of affect can disrupt negative affect. In A. Todorov, S. Fiske, D. Prentice (Eds.). *Social neuroscience: Toward understanding the underpinnings of the social mind* (pp. 188–209). Available from http://dx.doi.org/10.1093/acprof:oso/9780195316872.003.0013.

Lindquist, K. A., Wager, T. D., Kober, H., Bliss-Moreau, E., & Barrett, L. F. (2012). The brain basis of emotion: A meta-analytic review. *Brain and Behaviour Science, 35*, 121–202.

Chater, N., & Loewenstein, G. (2016). The under-appreciated drive for sense-making. *Journal of Economic Behavior & Organization, 126*(B), 137–154.

Lloyd, N. (1968). *The golden encyclopedia of music.* New York, NY: Golden Press.

Lord, C., Ross, L., & Lepper, M. (1979). Biased assimilation and attitude polarization: The effects of prior theories on subsequently considered evidence. *Journal of Personality and Social Psychology, 37*(11), 2098–2109.

Lord, C. G., Lepper, M. R., & Preston, E. (1984). Considering the opposite : A corrective strategy for social judgment. *Journal of Personality and Social Psychology, 47*(6), 1231–1243.

Lorenz, K. (1981). *The foundations of ethology.* New York, NY: Springer Verlag.

Lupasco, S. (1947). *Logique et contradiction.* P.U.F., Paris.

Luther M. (1538). *Preface to Symphoniae jucundae.* See W&T, p. 102.

Malevich, K. (1913). *Black Square, Oil on canvas, 41 3/4 × 41 7/8 in. (106.2 × 106.5 cm.), State Russian Museum, St. Petersburg. The defining work of Russian Suprematism movement, black square on white field.*

Martinez, J. L. (1981). *Endogenous peptides and learning and memory processes. behavioral biology.* New York, NY: Academic Press.

Maruskin, L. A., Thrash, T. M., & Elliot, A. J. (2012). The chills as a psychological construct : Content universe, factor structure, affective composition, elicitors, trait antecedents, and consequences. *Journal of Personality and Social Psychology, 103*(1), 135–157.

Masataka, N. (2006). Preference for consonance over dissonance by hearing newborns of deaf parents and of hearing parents. *Developmental Science, 9*(1), 46–50.

Masataka, N. (2008). The origins of language and the evolution of music: A comparative perspective. *Physics of Life Reviews, 6*(2009), 11–22.

Masataka, N., & Perlovsky, L. I. (2012a). Music can reduce cognitive dissonance. *Nature Precedings,* hdl:10101/npre.2012.7080.1; <http://precedings.nature.com/documents/7080/version/1>.

Masataka, N., & Perlovsky, L. I. (2012b). The efficacy of musical emotions provoked by Mozart's music for the reconciliation of cognitive dissonance. *Scientific Reports, 2.* Article number: 694. Available from http://dx.doi.org/10.1038/srep00694; <http://www.nature.com/srep/2013/130619/srep02028/full/srep02028.html>.

Masataka, N., & Perlovsky, L. I. (2013). Cognitive interference can be mitigated by consonant music and facilitated by dissonant music. *Scientific Reports, 3.* Article number: 2028 (2013). Available from http://dx.doi.org/10.1038/srep02028.

Mattheson, J. (1739). *The complete music master.* See W&T, p. 217.

Mayer, J. D., Salovey, P., Caruso, D. R., & Sitarenios, G. (2001). Emotional intelligence as a standard intelligence. *Emotion, 1*(3), 232–242.

Mayorga, R., & Perlovsky, L. I. (Eds.), (2008). *Sapient systems* London, UK: Springer.

McAllister, J. W. (1999). *Beauty and revolution in science* (p. 95). Ithaka, NY: Cornell Univ Press.

McCrae, R. R. (2007). Aesthetic chills as a universal marker of openness to experience. *Motivation and Emotion, 31*, 5–11. Available from http://dx.doi.org/10.1007/s11031-007-9053-1.

McCrae, R. R. (2009). The five-factor model of personality traits: Consensus and controversy. In P. J. Corr, & G. Matthews (Eds.), *The Cambridge handbook of personality psychology* (pp. 148–161). Cambridge, UK: Cambridge University Press.

McCrae, R. R., & Costa, Paul T., Jr. (1989). Reinterpreting the Myers-Briggs type indicator from the perspective of the five-factor model of personality. *Journal of Personality, 57* (1), 17–40. Available from http://dx.doi.org/10.1111/j.1467-6494.1989.tb00759.x.

McDermott, J. (2008). The evolution of music. *Nature, 453*, 287–288.

McDermott, J., & Houser, M. (2003). The origins of music: Innateness, uniqueness, and evolution. *Music Perception, 23*, 29–59.

Meehl, P. (1978). Theoretical risks and tabular asterisks: Sir Karl, Sir Ronald, and the slow progress of soft psychology. *Journal of Consulting and Clinical Psychology, 46*(4), 806–834.

Messing, R. B., Jensen, R. A., Martinez, J. L., Spiehler, V. R., Jr, Vasquez, B. J., Soumiereu-Mourat, B., ... McGaugh, J. L. (1979). Naloxone enhancement of memory. *Behavioral and Neural Biology, 27*, 266–275.

Meyer, L. B. (1956). *Emotion and meaning in music.* Chicago, IL: Chicago University Press.

Meyer, L. B. (1957). Meaning in music and information theory. *Journal of Aesthetics and Art Criticism, 15*(4), 412–424.

Meyer, R. K., Palmer, C., & Mazo, M. (1998). Affective and coherence responses to Russian laments. *Music Perception, 16*(1), 135–150.

Miller, G. F. (2000). Evolution of human music through sexual selection. In N. L. Wallin, B. Merker, & S. Brown (Eds.), *The origins of music* (pp. 329–360). Cambridge, MA: MIT Press.

Mithen, S. (2007). *The singing Neanderthals: The origins of music, language, mind, and body.* Cambridge, MA: Harvard University Press.

Mosing, M. A., Verweij, K. J. H., Madison, G., Pedersen, N. L., Zietsch, B. P., & Ullén, F. (2015). Did sexual selection shape human music? Testing predictions from the sexual selection hypothesis of music evolution using a large genetically informative sample of over 10,000 twins. *Evolution and Human Behavior, 3*, 359–366.

Nabokov, V. (1981). Lectures on Russian literature 1st edition, p.64, see Mariner Books, 2002.

Neuman, Y., Livshits, D., Perlovsky, L., & Cohen, Y. (2015). The personality of music genres. *Psychology of Music, 1-14*, 0305735615608526.

Nietzsche, F. (1876/1997). *Untimely meditations* (R.J. Hollingdale, Trans.). Cambridge, England: Cambridge University Press.

Nikolsky, A. (2015). Evolution of tonal organization in music mirrors symbolic representation of perceptual reality. Part-1: Prehistoric. *Frontiers in Psychology, 6*, 1405.

Nikolsky, A. (2016). Evolution of tonal organization in music optimizes neural mechanisms in symbolic encoding of perceptual reality. Part-2: Ancient to seventeenth century. *Frontiers in Psychology*, 7, 211.

Ortony, A., & Turner, T. J. (1990). What's basic about basic emotions? *Psychological Review, 97*, 315–331.

Ottosson, J. (2007). *The importance of nature in coping (Ph.D. thesis).* Sweden: Acta Universitatis Agriculurae Sueciae.

Panksepp, J. (1995). The emotional sources of "chills" induced by music. *Music Perception: An Interdisciplinary Journal, 13*(2), 171–207. Available from http://dx.doi.org/10.2307/40285693.

Panksepp, J., & Bernatzky, G. (2002). Emotional sounds and the brain: The neuro-affective foundations of musical appreciation. *Behavioural Processes, 60*, 133–155.

Panzarella, R. (1980). The phenomenology of aesthetic peak experiences. *Journal of Humanistic Psychology, 20*, 69–85.

Pascal, B. (1995). *Pensees.* London: Penguin books.

Patel, A. D. (2008). *Music, language, and the brain.* New York, NY: Oxford Univ. Press.

Patterson, T. A., Schulteis, G., Alvarado, M. C., Martinez, J. L., Bennett, E. L., Rosenzweig, M. R., & Hruby, V. J. (1989). Influence of opioid peptides on learning

and memory processes in the chick. *Behavioral Neuroscience, 103*(2), 429–437. Available from http://dx.doi.org/10.1037/0735-7044.103.2.429.
Paulucci, A., & Paulucci, H. (1962). Hegel on Tragedy. New York: Doubleday.
Pearson, D. G., & Tong, C. (2014). The great outdoors? exploring the mental health benefits of natural environments. *Frontiers in Psychology, 5*, 1178.
Perlovsky, L., & Kozma, R. (2007b). Neurodynamics of cognition and consciousness (editorial). In L. Perlovsky, & R. Kozma (Eds.), *Neurodynamics of cognition and consciousness*. Heidelberg, Germany: Springer Verlag.
Perlovsky, L. I. (2001a). *Neural networks and intellect*. New York, NY: Oxford Univ. Press.
Perlovsky, L. I. (2001b). *Mystery of sublime and mathematics of intelligence. Zvezda, 20018*, 174–190, St. Petersburg Russian.
Perlovsky, L. I. (2001c). *Neural networks and intellect: Using model-based concepts*. New York, NY: Oxford University Press, (3rd printing).
Perlovsky, L.I. (2002). Aesthetics and mathematical theory of intellect. Russian Academy of Sciences, Moscow, Russia: *Iskusstvoznanie, Journal of History and Theory of Art*, 2, 558–594 (in Russian).
Perlovsky, L. I. (2004). Integrating language and cognition. *IEEE Connections, Feature Article, 2*(2), 8–12.
Perlovsky, L. I. (2005). Evolution of consciousness and music. *Zvezda, 20058*, 192–223, St. Petersburg Russian.
Perlovsky, L. I. (2006a). *Joint evolution of cognition, consciousness, and music. Lectures in musicology, School of Music.* Columbus, OH: University of Ohio.
Perlovsky, L. I. (2006b). Music—The first principles. *Musical Theater*. <http://www.ceo.spb.ru/libretto/kon_lan/ogl.shtml> (14.12.01).
Perlovsky, L. I. (2006c). Toward physics of the mind: Concepts, emotions, consciousness, and symbols. *Physics of Life Reviews., 3*(1), 22–55.
Perlovsky, L. I. (2007a). Evolution of languages, consciousness, and cultures. *IEEE Computational Intelligence Magazine, 2*(3), 25–39.
Perlovsky, L. I. (2007b). Modelling field theory of higher cognitive functions. In A. Loula, R. Gudwin, & J. Queiroz (Eds.), *Artificial cognition systems* (pp. 64–105). Hershey, PA: Idea Group.
Perlovsky, L. I. (2007c). Neural dynamic logic of consciousness: KI. In L. I. Perlovsky, & R. Kozma (Eds.), *Neurodynamics of higher-level cognition and consciousness*. Heidelberg, Germany: Springer Verlag.
Perlovsky, L. I. (2007d). Symbols: Integrated cognition and language. In R. Gudwin, & J. Queiroz (Eds.), *Semiotics and intelligent systems development* (pp. 121–151). Hershey, PA: Idea Group.
Perlovsky, L. I. (2008a). Music and consciousness. *Leonardo, Journal of Arts, Sciences and Technology, 41*(4), 420–421.
Perlovsky, L. I. (2008b). Sapience, consciousness, and KI. Prolegomena to a physical theory. In R. Mayorga, & L. I. Perlovsky (Eds.), *Sapient systems*. London: Springer.
Perlovsky, L. I. (2009a). Language and cognition. *Neural Networks, 22*(3), 247–257. Available from http://dx.doi.org/10.1016/j.neunet.2009.03.007.
Perlovsky, L. I. (2009b). Language and emotions: Emotional Sapir-Whorf hypothesis. *Neural Networks, 22*(5–6), 518–526. Available from http://dx.doi.org/10.1016/j.neunet.2009.06.034.
Perlovsky, L. I. (2009c). 'Vague-to-Crisp' neural mechanism of perception. *IEEE Transactions on Neural Networks, 20*(8), 1363–1367.
Perlovsky, L. I. (2010a). Intersections of mathematical, cognitive, and aesthetic theories of mind. *Psychology of Aesthetics, Creativity, and the Arts, 4*(1), 11–17. Available from http://dx.doi.org/10.1037/a0018147.
Perlovsky, L. I. (2010b). The mind is not a Kludge. *Skeptic, 15*(3), 51–55.
Perlovsky, L. I. (2010c). Musical emotions: Functions, origin, evolution. *Physics of Life Reviews*, 7(1), 2–27.

Perlovsky, L. I. (2010d). Neural Mechanisms of the mind, Aristotle, Zadeh, and fMRI. *IEEE Transactions on Neural Networks, 21*(5), 718–733.

Perlovsky, L.I. (2010e). Science and Religion: *Scientific Understanding of Emotions of Religiously Sublime*, arXive.

Perlovsky, L. I. (2011a). Abstract concepts in language and cognition, Commentary on "Modeling the Cultural Evolution of Language" by Luc Steels. *Physics of Life Reviews, 8*(4), 375–376.

Perlovsky, L. I. (2011b). Emotions of "higher" cognition, Comment to Lindquist at al 'The brain basis of emotion: A meta-analytic review'. *Brain and Behaviour Sciences, 35*, 157–158, in print.

Perlovsky, L. I. (2011c). "High" cognitive emotions in language prosody. *Physics of Life Reviews, 8*(4), 408–409.

Perlovsky, L.I. (2011d). Language and cognition interaction: Neural mechanisms, computational intelligence and neuroscience. *Open Journal*, doi:10.1155/2011/454587. http://www.hindawi.com/journals/cin/contents/ (14/12/2011). Article ID 454587.

Perlovsky, L. I. (2011e). Language, emotions, and cultures: Emotional Sapir-Whorf hypothesis. *WebmedCentral Psychology 2011, 2*(2), WMC001580.

Perlovsky, L. I. (2011f). Music. Cognitive function, origin, and evolution of musical emotions. *WebmedCentral Psychology 2011, 2*(2), WMC001494.

Perlovsky, L. I. (2012a). Brain: Conscious and unconscious mechanisms of cognition, emotions, and language. *Brain Sciences, Special Issue "The Brain Knows More than It Admits", 2*(4), 790–834. <http://www.mdpi.com/2076-3425/2/4/790>.

Perlovsky, L. I. (2012b). The cognitive function of emotions of spiritually sublime. *Review of Psychology Frontier, 1*(1), 1–10. <www.j-psy.org>.

Perlovsky, L. I. (2012c). Cognitive function of music, Part I. *Interdisciplinary Science Reviews, 37*(2), 129–142.

Perlovsky, L. I. (2012d). Cognitive function, origin, and evolution of musical emotions. *Musicae Scientiae, 16*(2), 185–199. Available from http://dx.doi.org/10.1177/1029864912448327.

Perlovsky, L. I. (2012e). Emotions of "higher" cognition, Comment to Lindquist at al 'The brain basis of emotion: A meta-analytic review'. *Brain and Behaviour Sciences, 35* (3), 157–158.

Perlovsky, L. I. (2012f). Emotionality of languages affects evolution of cultures. *Review of Psychology Frontier, 1*(3), 1–13. <http://www.academicpub.org/rpf/paperInfo.aspx?ID=31>.

Perlovsky, L. I. (2013a). Language and cognition—Joint acquisition, dual hierarchy, and emotional prosody. *Frontiers in Behavioral Neuroscience*, 7, 123. Available from http://dx.doi.org/10.3389/fnbeh.2013.00123; <http://www.frontiersin.org/Behavioral_Neuroscience/10.3389/fnbeh.2013.00123/full>.

Perlovsky, L. I. (2013b). Learning in brain and machine—Complexity, Gödel, Aristotle. *Frontiers in Neurorobotics*. Available from http://dx.doi.org/10.3389/fnbot.2013.00023; <http://www.frontiersin.org/Neurorobotics/10.3389/fnbot.2013.00023/full>.

Perlovsky, L. I. (2014a). Aesthetic emotions, what are their cognitive functions? *Frontiers in Psychology, 5*, 98. <http://www.frontiersin.org/Journal/10.3389/fpsyg.2014.00098/full; Available from http://dx.doi.org/10.3389/fpsyg.2014.0009.

Perlovsky, L. I. (2014b). The cognitive function of music, Part II. *Interdisciplinary Science Reviews, 39*(2), 162–186.

Perlovsky, L. I. (2014c). Mystery in experimental psychology, how to measure aesthetic emotions? *Frontiers in Psychology, 5*, 1006. Available from http://dx.doi.org/10.3389/fpsyg.2014.01006.

Perlovsky, L. I. (2015a). Aesthetic emotions goals. Comment on "the Quartet theory of human emotions: An integrative and neurofunctional model" by Koelsch, Jacobs, Menninghaus, Liebal, Klann-Delius, von Scheve & Gebauer. *Physics of Life Reviews, 13*(2).

Perlovsky, L.I. (2015b). *How music helps resolve our deepest inner conflicts. The Conversation.*

Perlovsky, L. I. (2015c). Origin of music and the embodied cognition. *Frontiers in Psychology, 6*, 538–541. Available from http://dx.doi.org/10.3389/fpsyg.2015.00538.

Perlovsky, L. I., Bonniot-Cabanac, M.-C., & Cabanac, M. (2010). Curiosity and pleasure. *WebmedCentral Psychology 2010, 1*(12), WMC001275. <http://www.webmedcentral.com/article_view/1275>.

Perlovsky, L. I., Cabanac, A., Bonniot-Cabanac, M.-C., & Cabanac, M. (2013). Mozart effect, cognitive dissonance, and the pleasure of music. ArXiv 1209.4017. *Behavioural Brain Research, 244*, 9–14.

Perlovsky, L. I., Deming, R. W., & Ilin, R. (2011). *Emotional cognitive neural algorithms with engineering applications. Dynamic logic: From vague to crisp.* Heidelberg, Germany: Springer.

Perlovsky, L. I., & Ilin, R. (2010a). Grounded symbols in the brain, computational foundations for perceptual symbol system. *WebmedCentral Psychology 2010, 1*(12), WMC001357.

Perlovsky, L. I., & Ilin, R. (2010b). Neurally and mathematically motivated architecture for language and thought. Special issue "brain and language architectures: Where we are now?". *The Open Neuroimaging Journal, 4*, 70–80. <http://www.bentham.org/open/tonij/openaccess2.htm> (14.12.11).

Perlovsky, L. I., & Ilin, R. (2012). Mathematical model of grounded symbols: Perceptual symbol system. *Journal of Behavioral and Brain Science, 2*, 195–220. Available from http://dx.doi.org/10.4236/jbbs.2012.22024; <http://www.scirp.org/journal/jbbs/>.

Perlovsky, L. I., & Kozma, R. (Eds.), (2007a). *Neurodynamics of higher-level cognition and consciousness* Heidelberg, Germany: Springer-Verlag.

Perlovsky, L. I., & Levine, D. (2012). The drive for creativity and the escape from creativity: Neurocognitive mechanisms. *Cognitive Computation, 4*, 292–305. Available from http://dx.doi.org/10.1007/s12559-012-9154-3; <http://www.springerlink.com/content/517un26h46803055/>.

Perlovsky, L. I., & Mayorga, R. (2008). Preface. In R. Mayorga, & L. I. Perlovsky (Eds.), *Sapient systems*. London: Springer.

Perlovsky, L. I., & McManus, M. M. (1991). Maximum likelihood neural networks for sensor fusion and adaptive classification. *Neural Networks, 4*(1), 89–102.

Petrov, S., Fontanari, F., & Perlovsky, L. I. (2012). Categories of emotion names in web retrieved texts. *Cognitive Science Systems, 10*, 6–20.

Piff, P. K., Dietze, P., Feinberg, M., Stancato, D. M., & Keltner, D. (2015). Awe, the small self, and prosocial behavior. *Journal of Personality and Social Psychology, 108*(6), 883–899.

Pinker, S. (1997). *How the mind works.* New York, NY: Norton.

Plato (1997). Laws. In Tr. Sounders Cooper (Ed.), *Complete work* (pp. 700a–701b). Hackett, Cambridge, pp. 1388-1389.

Polkinghorne, D. E. (1988). *Narrative knowing and the human sciences.* Albany, NY: SUNY.

Poincare, H. (2001). *The value of science: Essential writings of Henri Poincare.* New York: Modern Library.

Polti, G. (1921). *The thirty-six dramatic situations.* Franklin, OH: J.K. Reeve.

Pope John XXII. (1323). *Bull Docta sanctorum.* See in W&T.

Popper, K. R. (1962). *Conjectures and refutations. The growth of scientific knowledge.* New York, NY: Basic Books.

Propp, Vladimir (1968). *Morphology of the folktale* (Vol. 9Austin, TX: University of Texas Press.

Purwins, H., Herrera, P., Grachten, M., Hazan, A., Marxer, R., & Serra, X. (2008a). Computational models of music perception and cognition I: The perceptual and cognitive processing chain. *Physics of Life Reviews, 5*, 151–168.

Purwins, H., Herrera, P., Grachten, M., Hazan, A., Marxer, R., & Serra, X. (2008b). Computational models of music perception and cognition II: Domain-specific music processing. *Physics of Life Reviews*, *5*, 169–182.

Rentfrow, P. J., & Gosling, S. D. (2003). The do re mi's of everyday life: The structure and personality correlates of music preferences. *Journal of Personality and Social Psychology*, *84*, 1236–1256.

Riasanovsky, N. V. (1992). *The emergence of romanticism*. New York, NY: Oxford University Press.

Roche, M. W. (2006). Introduction to Hegel's theory of tragedy. *Phaenex*, *1*(2), 11–20.

Russell, J. (1980). A circumplex model of affect. *Journal of Personality and Social Psychology*, *39*, 1161–1178.

Sadie, S. (Ed.), (1994). *Classical music pages, The Norton/Grove concise encyclopedia of music* New York, NY: W.W. Norton & Co. <http://w3.rz-berlin.mpg.de/cmp/musical_ history.html>.

Schaeffer, J.-M. (2015). *L'expérience esthétique. NRF essais*. Paris: Editions Gallimard.

Schlick, M., Mulder, H. L., & van de Velde-Schlick, B. F. B. (1979). *Philosophical papers* (Vol. 11) Dordrecht: D. Reidel Pub. Co.

Schoeller, F. (2015a). Knowledge, curiosity, and aesthetic chills. *Frontiers in Psychology*, *6*, 1546. Available from http://dx.doi.org/10.3389/fpsyg.2015.01546.

Schoeller, F. (2015b). The shivers of knowledge. *Human and Social Studies*, *Vol. 4* (Issue 3), 26–41. Available from http://dx.doi.org/10.1515/hssr-2015-0022.

Schoeller, F., & Perlovsky, L. I. (2015). Great expectations—Narratives and the elicitation of chills. *Psychology*, *6*(16), 2098–2102. Available from http://dx.doi.org/10.4236/ psych.2015.616205.

Schoeller, F., & Perlovsky, L. I. (2016). Knowledge-acquisition, meaning-making and aesthetic emotions: An experimental study of aesthetic chills. *Frontiers in Psychology*, Emotions, submitted, 02/05/2016.

Schoeller, F., & Perlovsky, L. (2016). Aesthetic chills: Knowledge-acquisition, meaning-making, and aesthetic emotions. *Frontiers in Psychology*, 7, 1093. doi: http://dx.doi. org/10.3389/fpsyg.2016.01093, PMCID: PMC4973431; http://www.ncbi.nlm.nih. gov/pmc/articles/PMC4973431/.

Schopenhauer, A. (1819, see 1998). *The World as Will and Idea*, (p.217), Everyman, Vermont.

Schulz, G. M., Varga, M., Jeffires, K., Ludlow, C. L., & Braun, A. R. (2005). Functional neuroanatomy of human vocalization: An H215O PET study. *Cerebral Cortex*, *1512*, 1835–1847.

Scruton, R. (1997). *The esthetics of music* (p. 343). Oxford, Great Britain: Oxford University Press.

Seyfarth, R. M., & Cheney, D. L. (2003). Meaning and emotion in animal vocalizations. *Annals of the New York Academy of Sciences*, *1000*, 32–55.

Simon, H. (1967). Motivational and emotional controls of cognition. *Psychological Review*, *Vol 74*(1), 29–39. Available from http://dx.doi.org/10.1037/h0024127.

Simonton, D. K. (1997). *Genius and creativity*. New York, NY: Ablex Publishing.

Sloboda, J. A. (1991). Music structure and emotional response: Some empirical findings. *Psychology of Music*, *19*(2), 110–120.

Sloboda, J. A., & Juslin, P. N. (2001). Psychological perspectives on music and emotion. In P. N. Juslin, & J. A. Sloboda (Eds.), *Music and emotion: Theory and research* (pp. 71–104). Oxford, GB: Oxford University Press.

Spelke, E. S., & Kinzler, K. D. (2007). Core knowledge. *Developmental Science*, *10*, 89–96.

Spinoza, B. (2005). *Ethics*. (E. Curley, Trans.). New York, NY: Penguin (Originally published in 1677).

Steels, L. (2011). Modeling the cultural evolution of language. *Physics of Life Reviews*, *8*(4), 339–356.

Steinbeis, N., Koelsch, S., & Sloboda, J. A. (2006). The role of harmonic expectancy violations in musical emotions: Evidence from subjective, physiological, and neural responses. *Journal of Cognitive Neuroscience*, *18*, 1380–1393.

Stravinsky, I. (1939–1940). *Poetics of music in the form of six lessons, bi-linugal edition* (A. Knodel, I. Dahl, Trans.). Cambridge, MA: Harvard Univ. Press. (Based on Stravinsky's Charles Eliot Norton lectures in Poetics at Harvard University).

Stravinsky, I. (1958). Conversations. In I. Stravinsky, & R. Craft (Eds.), *Conversations with Igor Stravinsky* (p. 1980). Berkely, CA: Univ. California Press.

Stroop, J. R. (1935). Studies of interference in serial verbal reactions. *Journal of Experimental Psychology*, *18*, 643–682.

Taruskin, R. (1995). *Text and act: Essays on music and performance*. New York & Oxford: Oxford Univ. Press.

Taruskin, R. (1997). *Defining Russia musically* (p. 322). Princeton, NJ: Princeton Univ. Press.

Taruskin, R. (1998). *The birth of contemporary Russia out of the spirit of Russian music*. <http://www.stanford.edu/group/Russia20> (18.04.16).

Tierno, M. (2002). *Aristotle's poetics for screenwriters: Storytelling secrets from the greatest minds in western civilization*. New York, NY: Hachette Book.

Thaler, R. H., & Sunstein, C. R. (2009). *Nudge*. New York, NY: Penguin.

Thompson, W. F, Schellenberg, E. G., & Husain, G. (2001). Arousal mood and the Mozart effect. *Psychological Science*, *12*(3), 248–251.

Tikhanoff, V., Fontanari, J. F., Cangelosi, A., & Perlovsky, L. I. (2006). *Language and cognition integration through modelling field theory: Category formation for symbol grounding, Book series in computer science* (Vol. 4131). Heidelberg: Springer.

Todorov, T. (1969). *Grammaire du Décaméron, Approaches to semiotics* (Vol. 3). The Hague: Mouton.

Trainor, L. (2004). Innateness, learning, and the difficulty of determining whether music is an evolutionary adaptation. *Music Perception*, *24*, 105–110.

Trainor, L. (2008). The neural roots of music. *Nature*, *453*(29), 598–599.

Trainor, L. (2015). The origins of music in auditory scene analysis and the roles of evolution and culture in musical creation. *Philosophical Transactions of the Royal Society B: Biological Sciences*, *370*(1664). <http://rstb.royalsocietypublishing.org/content/370/1664/20140089.full>.

Trainor, L., & Heinmiller, B. M. (1998). The development of evaluative responses to music: Infants prefer to listen to consonance over dissonance. *Infant Behavior and Development*, *21*, 77–88.

Trehub, S. E. (2003). The developmental origins of musicality. *Nature Neuroscience*, *6*(7), 669–673.

Trehub, S. E. (2008). Music as a dishonest signal. *Behavioural and Brain Sciences*, *31*, 598–599.

Trehub, S. E., Becker, J., & Morley, I. (2015). Cross-cultural perspectives on music and musicality. *Philosophical Transactions of the Royal Society B: Biological Sciences*, *370*, 20140096.

Trout, J. D. (2004). The philosophical legacy of Meehl (1978): Confirmation theory, theory quality, and scientific epistemology. *Applied & Preventive Psychology*, *11*, 73–76.

Tversky, A., & Kahneman, D. (1974). Judgment under uncertainty: Heuristics and biases. *Science*, *185*, 1124–1131.

Wallin, N., Merker, B., Brown, S. (2010) *The origins of music*. Cambridge, MA: MIT Press.

Weiss, P., & Taruskin, R. (1984a). *Music in the western world* (p. 15). New York, NY: Schirmer, Macmillan.
Wittkower, W. (1960). The changing concept of proportion. *Daedalus, Winter*, 199–215.
Wood, A., Lupyan, G., & Niedenthal, P. (2016). Why do we need emotion words in the first place? Commentary on Lakoff. *Emotion Review, 8*(3), 274–283.
Wuss, P. (2009). *Cinematic narration and its psychological impact: Functions of cognition, emotion and play*. Newcastle upon Tyne, UK: Cambridge Scholars.
Yardley, H., Perlovsky, L. I., & Bar, M. (2011). *Predictions and incongruency in object recognition: A cognitive neuroscience perspective. Detection and identification of rare audiovisual cues. studies in computational intelligence series*. New York, NY: Springer Publishing.
Zentner, M., Grandjean, D., & Scherer, K. R. (2008). Emotions evoked by the sound of music: Characterization, classification, and measurement. *Emotion, 8*, 494–521.
Zipf, G. K. (1949). *Human behaviour and the principle of least effort*. Oxford, England: Addison-Wesley Press.
Zuckerman, M., Eysenck, S. B. J., & Eysenck, H. J. (1978). Sensation seeking in England and America: Cross-cultural, age, and sex comparisons. *Journal of Consulting and Clinical Psychology, 46*(1), 139–149. Available from http://dx.doi.org/10.1037/0022-006X.46.1.139.

INDEX

Note: Page numbers followed by "*f*" and "*t*" refer to figures and tables, respectively.

D

E

F

N

O

P

T

Printed and bound by CPI Group (UK) Ltd, Croydon, CR0 4YY
08/06/2025
01896868-0008